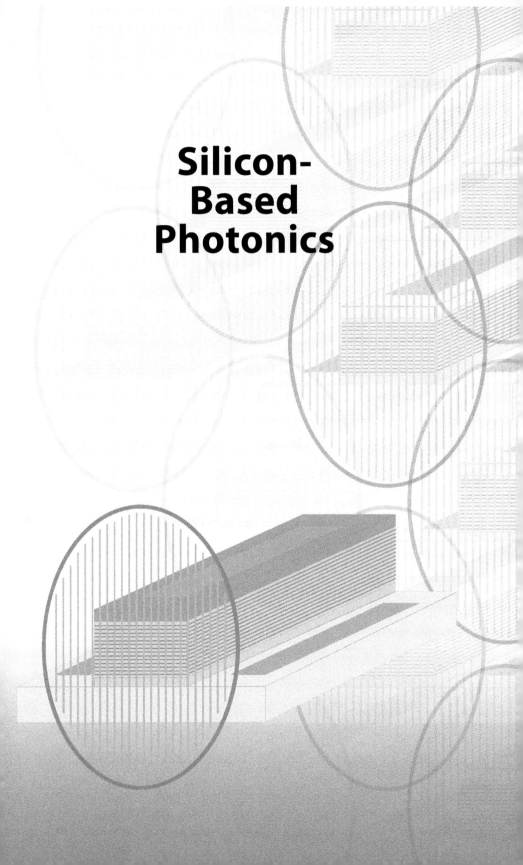

Silicon-Based Photonics

Silicon-Based Photonics

Erich Kasper
Jinzhong Yu

Jenny Stanford
Publishing

Published by

Jenny Stanford Publishing Pte. Ltd.
Level 34, Centennial Tower
3 Temasek Avenue
Singapore 039190

Email: editorial@jennystanford.com
Web: www.jennystanford.com

British Library Cataloguing-in-Publication Data
A catalogue record for this book is available from the British Library.

Silicon-Based Photonics
Copyright © 2020 by Jenny Stanford Publishing Pte. Ltd.

All rights reserved. This book, or parts thereof, may not be reproduced in any form or by any means, electronic or mechanical, including photocopying, recording or any information storage and retrieval system now known or to be invented, without written permission from the publisher.

For photocopying of material in this volume, please pay a copying fee through the Copyright Clearance Center, Inc., 222 Rosewood Drive, Danvers, MA 01923, USA. In this case permission to photocopy is not required from the publisher.

ISBN 978-981-4303-24-8 (Hardcover)
ISBN 978-981-4303-25-5 (eBook)

Contents

Preface	xi

1. Introduction — **1**
 1.1 Si Photonics — 2
 1.2 Si-Based Photonics — 4
 1.3 Book Content — 6

2. Band Structure and Optical Properties — **9**
 2.1 Bonding Lengths in a Diamond/Zincblende Lattice — 10
 2.2 Dielectric Function — 12
 2.3 Absorption Processes — 15
 2.4 Direct Group IV Semiconductors — 17

3. Planar Waveguides — **21**
 3.1 Modes in the Slab Waveguide — 22
 3.2 Strip Waveguides and Rib Waveguides — 31
 3.3 Loss in a Silicon Optical Waveguide — 39
 3.4 Polarization Dependence of Silicon Waveguides — 44
 3.5 Summary — 46

4. Microring Resonators — **49**
 4.1 Introduction — 49
 4.2 Principle of the Microring Resonator — 50
  4.2.1 Single Microring — 50
  4.2.2 Cascaded Microrings — 52
   4.2.2.1 Transfer matrix units — 53
   4.2.2.2 Series-coupled microring resonators — 55
   4.2.2.3 Parallel-coupled microring resonators — 56
 4.3 Optical Properties of Microring Resonators — 58
  4.3.1 Properties of Amplitude — 58
   4.3.1.1 Single microring — 58

		4.3.1.2 Cascaded microrings	61
	4.3.2	Properties of Phase	64
4.4	Design of Microring Resonators		68
	4.4.1	Waveguide Design	68
	4.4.2	Coupler Design	69
	4.4.3	Internal Loss	72
4.5	Fabrication and Measurement of Microring Resonators		73
	4.5.1	Fabrication Flow	74
	4.5.2	Electron Beam Lithography	75
	4.5.3	Dry Etching Process	76
	4.5.4	Silicon Oxide Growing	78
	4.5.5	Measurement	79
4.6	Applications of Microring Resonators		80
	4.6.1	Optical Filter	81
	4.6.2	Nonlinear Optical Devices	85
	4.6.3	Optical Buffer	87
4.7	Summary		88

5. Optical Couplers — **95**

5.1	Introduction		95
5.2	Spot-Size Converters		98
	5.2.1	Tapered Spot-Size Converter	99
	5.2.2	Inverted Tapered Spot-Size Converter	102
	5.2.3	Slot-Waveguide Spot-Size Converter	104
5.3	Prism Couplers		105
	5.3.1	Inverted Prism Coupler	105
	5.3.2	Graded Index Half-Prism Coupler	106
5.4	Grating Couplers		108
	5.4.1	Grating Coupler with a Vertical Coupling Structure	109
	5.4.2	Horizontal Dual Grating–Assisted Coupler	115
5.5	Conclusion		116

6. Photonic Crystals — **123**

6.1	Introduction		123

6.2		Master Equation	125
6.3		Calculation Methods	126
	6.3.1	PWE Method	126
	6.3.2	FDTD Method	127
6.4		Silicon-Based PC Slab	129
	6.4.1	Important Points about the SOI PC Slab	129
	6.4.2	Fabrication of Silicon-Based PC Slab	131
6.5		SOI PC Devices	133
	6.5.1	SOI PC Waveguides	133
	6.5.2	SOI PC Microcavities	135
	6.5.3	SOI PC Filters	138
6.6		Conclusions	142

7. Slow Light in a Silicon-Based Waveguide   **145**

7.1		Introduction	145
7.2		Concept	146
7.3		Slow Light in Microring Resonator Waveguides	147
	7.3.1	Single-Microring Resonator	149
	7.3.2	SCISSOR Configuration Microring Resonators	151
	7.3.3	CROW Configuration Microring Resonators	153
	7.3.4	Experimental Progress	154
7.4		Slow Light in Photonic Crystals	156
	7.4.1	Generation of Slow Light in a Photonic Crystal Waveguide	156
	7.4.2	Experimental Verification of Slow Light in Photonic Crystals	158
7.5		Conclusion	161

8. Light Emitters   **167**

8.1		Bandgap Emission Mechanisms	169
	8.1.1	Indirect Semiconductor Transitions	169
	8.1.2	Brillouin Zone Folding from the Superlattice	171
	8.1.3	Localization of Wave Functions by Quantum Structures and Porous Si	172

viii | *Contents*

8.2	Germanium Light-Emitting Diodes	173
	8.2.1 Competition between Direct and Indirect Transitions	173
	8.2.2 Influence of Doping, Strain, and Sn Alloying	175

9. Detectors — **179**

9.1	Detection Principles	179
9.2	Detector Configuration and Wavelength Considerations	181
	9.2.1 Vertical Incidence Detection	181
	9.2.2 Lateral Incidence Detection	182
	9.2.3 Wavelength Selection	183
9.3	Photon Detector Structure	190
	9.3.1 P/N Junction Photodiode	191
	9.3.2 Schottky Photodiode	195
	9.3.3 Avalanche Photodetector	198
9.4	Spectral Range	200
	9.4.1 UV Detectors	201
	9.4.2 Visible Spectrum	201
	9.4.3 NIR Spectrum	202
	9.4.4 SiGe/Si Quantum Dot (Well) IR Photodetectors	211
9.5	High-Speed Operation	212
9.6	Outlook	216

10. Modulators — **225**

10.1	Optical Modulation	227
10.2	Modulation Mechanisms	229
	10.2.1 Electro-optic Effect	229
	10.2.2 Thermo-optic Effect	231
	10.2.3 Plasma Dispersion Effect	232
	10.2.4 Strain and Quantization Effects	235
	10.2.4.1 Strain-induced linear electro-optic effect	235
	10.2.4.2 Quantum-confined Stark effect	235
10.3	Structures for the SOI Waveguide Modulator	236
	10.3.1 Electrical Structures	236

		10.3.1.1 Forward-biased p-i-n diode structure	236
		10.3.1.2 Reverse-biased p/n junction structure	238
		10.3.1.3 MOS capacitor structure	239
	10.3.2	Optical Structures	241
		10.3.2.1 Mach–Zehnder interferometer structure	241
		10.3.2.2 Fabry–Perot resonator structure	243
		10.3.2.3 Microring resonator structure	244
10.4	Summary		247

11. Extension of the Wavelength Regime — 251

11.1	Germanium Tin	251
	11.1.1 Bandgap of GeSn: Indirect to Direct Transition	252
	11.1.2 Epitaxial Layer Structures on Silicon	254
	11.1.3 Optical and Electrical Results	256
	11.1.4 Material Science Challenges	261
	11.1.5 Ternary Silicon Germanium Tin Alloys	262
11.2	Tensile-Strained Germanium	263
11.3	Conclusions	266

12. Laser — 269

12.1	Silicon-Based Laser Approaches	271
12.2	Basic Laser Physics	273
	12.2.1 Spontaneous Emission in a Two-Level System	273
12.3	Optical Gain in a Semiconductor	275
12.4	Semiconductor Heterostructure Laser	280
12.5	Conduction Band Occupation in an Indirect Semiconductor	284
	12.5.1 Steady-State Occupation Model for Indirect Semiconductors	285
	12.5.2 Effective Density of State for Ge	288
12.6	Influence of Strain and Sn Alloying	292
	12.6.1 Injected Carrier Densities and Layer Doping	296

		12.6.2 Internal Quantum Efficiency and Threshold Current	299
	12.7	Experiments	303
		12.7.1 Experimental Verification of Laser Light Emission from Group IV Materials	308
	12.8	Summary and Outlook	310
13.	**Future Challenges**		**317**
	13.1	Group IV Laser Performance	318
	13.2	Monolithic Integration Issues	320
Index			327

Preface

Both authors were guest professors at the same time at the Huazhong University of Science and Technology (HUST) in Wuhan, China. The Yangtse River area around Wuhan is known in China as Optics Valley, with broad research and manufacturing activities in fiberglass, optical sensors, optoelectronic devices, and photonic systems. We were in touch before on different topics of semiconductor device physics and material science but focused during our stay in Wuhan on silicon photonics.

Silicon photonics evolved rapidly as a research topic with enormous application potential. Essentially, the following three phases were responsible for continuous progress in this research area. The large refractive index contrast of silicon-on-insulator (SOI) waveguides allowed densely packed and strongly curved waveguides on chip-size carriers in the first phase. The second phase benefitted from progress in silicon microelectronics, with high-performance foundry service available to design and system groups. In this second phase, the number of involved groups jumped up and the economic fabrication sparked the vision of integration of photonic and microelectronic systems. A fundamental problem remained: the waveguide has to be transparent in the used infrared wavelength regime, but for conversion of the electrical signal in an optical one and vice versa, a strong light-matter interaction is requested. In the third phase, this photonics problem is tackled with semiconductor heterostructure devices where the heterostructure partner offers a higher cutoff wavelength as silicon.

We felt that the settled first two phases are well treated in monographs and edited books on silicon photonics whereas the reader has to consult proceeding volumes to monitor the rapid progress of the third phase.

We tried in the given book, titled *Silicon-Based Photonics*, to merge a concise treatment of classical silicon photonics with the description of principles, prospects, challenges, and technical solution paths of adding silicon-based heterostructures.

We acknowledge the help of many colleagues with whom we discussed different topics at common conference meetings and at personal meetings. Substantial contributions to the design of text and graphics of various chapters were made by Michael Oehme, Xuejun Xu, Kang Xiong, Qingzhong Huang, Xi Xiao, Yingtao Hu, Yuntao Li, Hejun Yu, Zhiyong Li, Hao Xu, and Yu Zhu.

We thank K. Ye, A. Raju, M. Kaschel, M. Schmidt for valuable help with preparation and critical reading. The friendly team of the publisher supported continuously the formal aspects of the work.

Erich Kasper

Jinzhong Yu

April 2020

Chapter 1

Introduction

Who are the technical parents of silicon photonics? Undoubtedly, the long-lasting technological success of silicon microelectronics, with their high integration levels on one side and the complete replacement of long-distance wire communication by optical fiber glass transmission on the other side, nurtured the demand to join optical waveguide transmission and reliable system integration on a silicon wafer scale. The pioneers (e.g., Soref and Lorenzo [1], Petermann [2], and Abstreiter [3]) proposed in the 1980s and the early 1990s waveguide and optoelectronic device integration on silicon although the indirect semiconductor silicon was considered as less favorable for optical functions. Indeed, the device and circuit development in microelectronics and optoelectronics started in different directions. Focus [4] on a basically simple device type (metal oxide semiconductor transistor) and on a few materials for the technology (semiconductor Si, dielectrics SiO_2, metal Al) made circuit design and integration in microelectronics easy. Progress in device and circuit performance was achieved by shrinkage of the device dimensions. This period of microelectronics, named "dimension scaling," lasted from the beginning of integrated circuit manufacturing (around the year 1975) to about the years 2000–2005. In optoelectronics, the performance was driven by sophisticated heterostructures based on III/V semiconductors that had excellent absorption, emission, and modulation properties for

---

Silicon-Based Photonics
Erich Kasper and Jinzhong Yu
Copyright © 2020 Jenny Stanford Publishing Pte. Ltd.
ISBN 978-981-4303-24-8 (Hardcover), 978-981-4303-25-5 (eBook)
www.jennystanford.com

optical signals. At this time (around the year 2000) it was clear that telecommunication was governed at far distances (more than 10 km) by optical signal transmission and at near distances by electrical signals. Near distances meant distances within an enterprise (0.1–10 km), between racks (1–100 m), between boards (10–100 cm), between chips (1–10 cm), and on a chip (<1 cm). With increasing speed (3–10 GHz), the optical communication was predicted to be competitive also at near distances because electrical connections suffer from the resistance-capacitance (RC) limitation of the speed of interconnects. Irrespective of the speed of devices, the speed of electronic circuits is ultimately limited by the interconnect time delay that is inversely proportional to the RC product.

1.1 Si Photonics

Silicon (Si) photonics is the prime candidate to cover the application range at the border region between pure electrical and pure optical solutions. The essential property of planar Si waveguides is due to their high refractive index contrast when fabricated on silicon-on-insulator (SOI) substrates (Fig. 1.1). These commercially available SOI substrates are composed of a Si wafer with a thin (typically 1 µm) SiO_2 (insulator) layer and an even thinner (typically 0.2–0.4 µm) Si layer on top.

The waveguide formation along a lithographically defined pattern uses etching (mainly dry etching by a reactive gas). Ridge waveguides show in cross section a ridge (Fig. 1.1, left side) on a partially etched Si top layer surrounding. The waveguide of small dimensions is called a nanowire if the complete Si layer outside the waveguide is etched (Fig. 1.1, right side). The waveguide core from a semiconductor like Si has a high refractive index n (about 3.5). The cladding from glass has a much lower refractive index, of about 1.5. Figure 1.1 does not show the top cover from glass that is used for circuit protection. Table 1.1 compares the index contrast $\Delta n/n$ for different waveguide materials.

The refractive index contrast of 60% from SOI waveguides is much higher than in fiberglass (typically 1%) but also substantially higher than in insulator (SiO_2/SiN) or semiconductor heterostructure (e.g., SiGe/Si) waveguides (typically 10%). This high index contrast allows single-mode waveguides of small dimensions, high curvature of waveguides lines, and high packing density. The SOI waveguide preparation with standard technologies of Si microelectronics guarantees high yield and reproducibility, favors cost-effective fabrication, and offers monolithic integration with electronic supply and readout circuits. The availability of the silicon foundry service [5] gives manifold options for fables activities in design, characterization, and system applications. The silicon waveguide is absorbing in the visible spectral range but gets transparent in the infrared (wavelengths $\lambda > 1.2$ µm). Especially, the telecommunication wavelength regimes around 1.3 µm and 1.55 µm are compatible with Si photonics [6, 7]. However, the available wavelength regime in low-doped Si is much broader, at least up to 8 µm. Impurities like oxygen or carbon may disturb the transparency in wavelength bands above 8 µm [8].

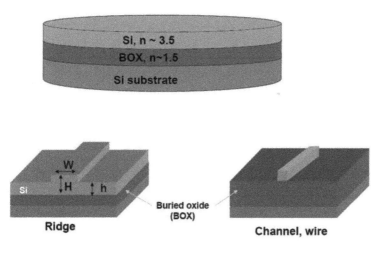

Figure 1.1 Silicon-on-insulator (SOI) waveguide structure. Upper part: Starting substrate consisting of a Si substrate covered with an oxide layer (buried oxide [BOX]) and a single crystalline Si top layer. After selective etching, a waveguide with a width W and a height H is formed (lower part). Partial etching of the top Si layer down to a thickness h creates a ridge waveguide (left side), and complete etching forms a wire structure. A protective oxide layer (not shown here) covers the waveguide structure.

4 | *Introduction*

Table 1.1 Refractive index contrast of typical waveguide structures

Material	n	$\Delta n/n$
Glass fiber	1.5	1%
SiO_2/SiN	1.5	10%
Semiconductor heterostructure	3.5	10%
SOI	3.5	60%

Note: Given is the refractive index n of the core material and the relative change $\Delta n/n$ between the core and the cladding. The high index contrast of SOI waveguides allows dense packing of photonic structures.

1.2 Si-Based Photonics

Active devices in photonics for emission, detection, and absorption modulation need strong light-matter interaction in the selected infrared transmission regime. This needs a different material for the active devices that should provide a strong light-matter interaction at the selected wavelength regime but with a good transparency of the Si waveguides.

As favorite solutions for these contrary requests for waveguide and active device emerged heterostructure semiconductor waveguide/device combinations. The semiconductor for the device needs to be chosen with a bandgap E_g lower than that of Si ($E_{gSi} = 1.12$ eV at room temperature). This provides a wavelength regime

$$1.24 \, \mu m/E_g \, (eV) < \lambda > 1.1 \, \mu m \tag{1.1}$$

that fulfils both high transparency of the Si waveguide and strong light-matter interaction for small active devices.

Silicon-Based Photonics as the title of this book refers to the fact that a silicon-based heterostructure is essential to fulfilling all the basic photonic system functions. The heterostructures are chosen either from a group III/V semiconductor on Si or preferably from a group IV semiconductor on Si. In the latter case, "group IV photonics" is an alternative term for this flourishing research topic.

The envisaged application spectrum of Si photonics started with telecommunication [9] networks in metro areas and then expanded to data center and cloud service [10], to on-chip clock distribution, and to links between cores of multicore processors and links between core and memories. The last-mentioned application spectra address the bottleneck in on-chip communication in ultra-large-scale integrated circuits. Figure 1.2 shows the principally simple arrangement of the optical path on the example of the clock distribution. The laser light couples with the chip from an external laser via a fixed monomode fiberglass mounting. The use of an external laser has the advantages that available high-performance lasers may be used, that power for the laser operation is distributed outside the thermally stressed electronic chip, and that this external clock laser may be used for several chips. The light is modulated either externally or internally, and then it is distributed via waveguides into different subchip areas. In Fig. 1.2, an internal modulator varies the laser light intensity that is distributed to different chip regions by the waveguide lines split up from the input line. Similar schemes provide fast access to memory contents for processor operation. Germanium photodetectors (d) convert the optical signal into an electrical signal at the waveguide line ends. The electrical signal is then distributed around within the small chip regions without the power and speed problems of the electrical clock distribution on large chip areas. Although the scheme is rather simple, the realization on an already complex electronic chip needs to overcome integration process and packaging issues. The majority of system investigations [11] use the common telecom wavelength of 1.3–1.6 µm, but years ago Soref [12] had already backed the use of the broad available wavelength spectrum from near infrared to mid-infrared. This broad spectrum not only delivers an incredible number of frequency slots for communication but also offers new possibilities for imaging and sensing in biology and chemistry [9, 11, 13]. The wavelength regime between 1.5 µm and 3 µm spans a frequency regime from 200 THz to 100 THz, a difference that is a thousand times the frequency span now used or planned for mobile communication or automotive radar. Interference of two lasers with slightly different wavelengths provides a signal the frequency of which may be chosen from the millimeter wavelength to the terahertz range, depending on the frequency difference [14]. Compact chip-sized solutions based on silicon photonics will certainly push the applications in this area between microwave and optics.

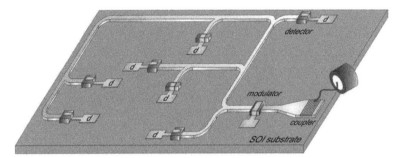

Figure 1.2 Scheme of an optical clock frequency distribution. Laser light from an external source is coupled in the chip waveguide via a grating coupler structure (from the right side).

1.3 Book Content

The book focuses on the integration of heterostructure devices with Si photonics, which has resulted in the recent advances of a photonics branch that is named "Si-based photonics." Some material and device phenomena appear in heterostructures not present in Si stand-alone technology. The most prominent of these phenomena are lattice mismatch, dislocation generation, high elastic strain, and metastable alloys. Strong progress in material science now allows the utilization of the effects from strain and metastability. These effects are also interesting in microelectronics for performance improvements, but in Si-based photonics, they are necessary for full functionality.

The reader will find basics about band structure and optical properties in Chapter 2. A concise treatment of classical Si-photonics topics is given in Chapters 3–7, with explanations of waveguides, resonators, couplers, photonic crystal, and slow light. These chapters are mainly for readers who are interested in the topic because of its increasing importance in different fields. Chapters 8–12 cover different device structures for light emission, detection, modulation, extension of the wavelength beyond 1.6 µm, and lasing. An outlook for the next challenges is given in Chapter 13.

With the given mix of basics and front-end research, the authors hope to meet the demands of a broad audience, from students and researchers to engineers who are already working in this field or who intend to start studies because of the widespread potential in technology and application.

References

1. R. A. Soref and J. P. Lorenzo (1986). All-silicon active and passive guided wave components for $\lambda = 1.3$ μm and 1.6 μm, *IEEE J. Quantum Electron.*, **22**, 873–879.

2. A. Splett, J. Schmidtchen, B. Schuppert, K. Petermann, E. Kasper and H. Kibbel (1990). Low loss optical ridge waveguides in a strained GeSi epitaxial layer grown on silicon, *Electron. Lett.*, **26**, 1035.

3. G. Abstreiter (1992). Engineering the future of electronics, *Phys. World*, **5**, 36.

4. E. Kasper, H. J. Mussing and H. G. Grimmeiss (eds.) (2009). *Advances in Electronic Materials* (TransTech Publication, Zurich, Switzerland).

5. L. Chrostowski and M. Hochberg (2015). *Silicon Photonics Design* (Cambridge Univ. Press) ISBN: 978-1-107-08545-9

6. G. T. Read (2008). *Silicon Photonics: The State of the Art* (John Wiley, Hoboken) ISBN: 978-0-470-02579-6.

7. G. T. Read and A. P. Knights (2004). *Silicon Photonics: An Introduction* (John Wiley, Hobokon) ISBN: 978-0-470-87034-7.

8. Focus Issue on Silicon Photonics (2010). *Nat. Photonics*, **4**(8), 491.

9. S. Fathpour and B. Jalali (eds.) (2012). *Silicon Photonics for Telecommunication and Biomedicine* (CRC Press, USA) ISBN: 978-1-4398-0637-1.

10. D. Inniss and R. Rubenstein (2017). *Silicon Photonics: Fueling the Next Information Revelation* (Elsevier, Morgan Kaufmann) ISBN: 978-0-1280-2975-6

11. L. Vivien and L. Pavesi (eds.) (2013). *Handbook of Silicon Photonics* (CRC Press, USA) ISBN: 978-1-4398-3610-1.

12. R. A. Soref, S. J. Emelett and W. R. Buchwald (2006). Silicon waveguided components for the long-wave infrared region, *J. Opt. A: Pure Appl. Opt.*, **8**, 840–848.

13. L. Pavesi and D. Lockwood (eds.) (2016). *Silicon Photonics III-Systems and Aplications* (Springer, Germany) Topics in Applied Physics, Vol. 122.

14. Focus Issue on Microwave Photonics (2015). *IEEE Microwave Mag.*, **16**(8), 28–94.

Chapter 2

Band Structure and Optical Properties

The band structure of semiconductors is mainly influenced by the lattice type, the bonding distance, and the ionicity of bonding partners. Have a look along the group IV element column of the periodic table, and one will find a thermodynamically stable diamond (zincblende) lattice type for silicon and a zincblende lattice type for AlP. The diamond (zincblende) lattice can be described by a cubic cell with eight atom positions, where all the positions are occupied by one element, A, (diamond lattice) or by two elements, A and B, from group III and group V elements, respectively. The lattice is strongly bonded by covalent forces between an atom and its immediate four neighbors (diamond lattice) and additionally by ionic forces in the zincblende lattice. The A (111) stacking of planes in the zincblende lattice may be described by an (abc) sequence a, b, c, a, b, c, ... This lattice scheme is only challenged at the low atomic number Z end (carbon, Z = 6; SiC, Z = 10) and at the high atomic number end (tin, Z = 50). For carbon the diamond lattice is metastable whereas the stable lattice is the layered graphite structure. A single or a few layers of graphite are now in research focus as graphene that can be considered as a narrow graphite quantum well. On the tin side the semiconducting α-Sn (gray tin) is only stable below 17°C; above room temperature the metallic β-Sn is stable. Alloying of Sn with small amounts of Ge stabilizes the diamond structure. Increasing ionicity by choosing atomic partners from group II (A) and group VI (B) favors a different lattice cell (hexagonal wurtzite lattice) that has

---

Silicon-Based Photonics
Erich Kasper and Jinzhong Yu
Copyright © 2020 Jenny Stanford Publishing Pte. Ltd.
ISBN 978-981-4303-24-8 (Hardcover), 978-981-4303-25-5 (eBook)
www.jennystanford.com

the same nearest-neighbor bonding as the zincblende lattice but a different stacking in the next nearest (111) plane. The (111) stacking in the hexagonal wurtzite lattice has an (ab) sequence with a, b, a, b, ... A wrong sequence is called a stacking fault, in the sense that the wurtzite lattice is considered as a zincblende lattice with a periodic stacking fault arrangement. A special situation is given with silicon carbide (SiC) where the stacking fault energy is very low, allowing a variety of (111) arrangements (polytypic SiC) from zincblende to wurtzite, with stacking faults of decreasing periods. The diamond lattice SiC is named "cubic-SiC" or "β-SiC."

2.1 Bonding Lengths in a Diamond/Zincblende Lattice

To simplify the discussion we concentrate now on the majority of applications with diamond (zincblende) lattices. The lattice constant a_0 increases with an increasing atomic number Z (mean number $Z = 1/2(Z_A + Z_B)$ for AB compounds). In Table 2.1 a comparison of group IV and group III/V elements/compounds is given. The lattice constant a_0 [1] and the normalized lattice constant $a_0/Z^{1/3}$ are shown in picometers (1 pm = 10^{-12} m). The atomic diameter increases roughly with $Z^{1/3}$.

The bonding length, that is the lattice constant a_0, determines largely the bandgap, with sharply decreasing gaps as the lattice constant increases. The bonding length is given as $a_0 \cdot \sqrt{3}/4$. The direct bandgap that separates the valence band maximum (which is always at the Brillouin zone center, wave vector $k = 0$, named Γ) from the lowest conduction band at Γ is given as E_{gdir}. From C to SiC to Si to Ge to Sn it reduces from 6.5 eV to 6 eV to 3.2 eV to 0.8 eV to 0 eV, spanning the transition from a semi-insulator to a semimetal.

In many semiconductors of this group the lowest bandgap is indirect, which means the lowest conduction band is either in (111) direction (called L point) or in (100) direction (Δ point; the final (100) in the Brillouin zone is named the X point). The indirect transition needs a phonon help to fulfil the momentum conservation. Absorption and photon emission are much weaker in indirect transitions than in direct transitions. In Fig. 2.1 the indirect bandgaps are named E_{gL} (Γ valence band to L conduction band transition) or E_{gx} (Γ valence band to X or Δ conduction band transition). The L conduction band energies decrease rapidly, with bond length dominating the indirect bandgap material Ge, whereas the indirect semiconductors C and Si are characterized by a Δ (E_{gx}) conduction band minimum.

Table 2.1 Cubic diamond/zincblende lattice

Element/ compound	Atomic number Z	Lattice constant a_0 (pm)	Normalized lattice constant $a_0/Z^{1/3}$ (pm)
C (diamond)	6	356.7	196.3
BN	6	361.15	198.7
SiC	10	436	202.4
AlN	10	438	203.3
Si	14	543.1	225.3
AlP	14	546.7	226.8
Ge	32	564.6	177.8
GaAs	32	565.3	178.1
Sn	50	648.9	176.1
InSb	50	648	175.9

Note: Given are the lattice constant a_0 and the normalized lattice constant $a_0/Z^{1/3}$ (Z means atomic number) as a function of Z.

Table 2.2 Bandgaps E_{gdir}, E_{gL}, and E_{gx} of group IV elements and their III/V counterparts with the same mean atomic number Z

Element	Lattice type	Z	E_{gdir} (eV)	E_{gL} (eV)	E_{gx} (eV)
α-Sn	D	50	−0.41	0.14	0.9
InSb	ZB	50	0.25	1.08	1.71
Ge	D	32	0.8	0.66	0.85
GaAs	ZB	32	1.42	1.25	1.94
Si	D	14	3.2	1.65	1.12
AlP	ZB	14	3.62	—	2.49
SiC	ZB	10	6	4.2	2.2
AlN	ZB	10	6.2	—	—
C	D	6	6.5	9.2	5.45
BN	ZB	6	8.2	—	6.4

Note: The lattice type is diamond (D) or zincblende (ZB). Direct semiconductors are underlined, and the lowest transition is underlined.

Band Structure and Optical Properties

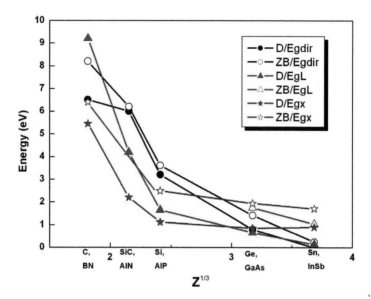

Figure 2.1 Band edge energies for different group VI and group III/V compounds as a function of $Z^{1/3}$ (Z means atomic number). Group IV is given by full symbols and group III/V by empty ones. E_g is black, E_{gL} red, and E_{gx} blue.

Compare group VI with group III/V materials to identify the influence of the ionicity that contributes to group III/V compounds. At all energy levels—direct and indirect, L or X—the ionicity increases the transition energies and strengthens the tendency for direct semiconductors; see AlN, GaAs, and InSb (Table 2.2) compared to Si, Ge, and Sn.

2.2 Dielectric Function

The macroscopic influence of a material on the electrical properties is described by a complex dielectric function $\underline{\varepsilon}_r$ [2],

$$\underline{\varepsilon}_r = \varepsilon_r' + j\varepsilon_r'', \tag{2.1}$$

that links the electric field strength $\vec{E}$ with the electric displacement vector $\vec{D}$ by the well-known relation

$$\vec{D} = \underline{\varepsilon}_r \varepsilon_0 \cdot \vec{E}. \tag{2.2}$$

For static fields or low-frequency modulation the displacement vector is in phase with the electric field and then the dielectric function reduces to the dielectric constant (permittivity) ε_r. For a higher circular frequency the dielectric displacement is out of phase, which is described by the complex dielectric function $\underline{\varepsilon}_r$. Figure 2.2 shows the dielectric function of silicon at room temperature in the spectral range of interband transitions as obtained by spectroscopic ellipsometry [3]. The dominant contribution to the frequency dependence of the dielectric function of semiconductors arises from electronic interband transitions between occupied valence band and empty conduction band states. Conservation of momentum imposes direct transitions without a change of wave vector for a first-order process.

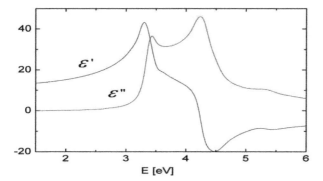

Figure 2.2 Dielectric function $\underline{\varepsilon} = \varepsilon' + j\varepsilon''$ of silicon versus photon energy.

The phase delay is explained by atomistic theories that take into account the polarization $\vec{P}$ of a material under the influence of an electrical field. One can immediately imagine that an atom is polarized by the different forces on the positive nucleus charge and the negative electron cloud charge or that in an ionic crystal an additional term is created by the different charges on the crystal lattice sites. The originally classical mechanics and later quantum mechanical theories of polarization $\vec{P}$ are the base of a quantitative understanding of the dielectric function of different dielectrics and semiconductors,

$$\vec{D} = \varepsilon_0 \cdot \vec{E} + \vec{P} \tag{2.3}$$

From Eqs. 2.2 and 2.3

$$\vec{P} = \varepsilon_0 (\underline{\varepsilon}_r - 1) \cdot \vec{E}. \tag{2.4}$$

The linear polarization property $(\varepsilon_r - 1)$ is termed electrical susceptibility χ:

$$\chi = \varepsilon_r - 1 \qquad (2.5)$$

For the description of an electromagnetic wave in a nonmagnetic material a related set of optical constants is more practical.

Let us define a complex quantity

$$\underline{n} = (\underline{\varepsilon_r})^{1/2}, \qquad (2.6)$$

which is the square root of the dielectric function.

This complex quantity n is composed of the refractive index n and the absorption index κ:

$$\underline{n} = n - j\kappa \qquad (2.7)$$

This follows from the solution (Maxwell's equations) of an electromagnetic wave in a nonmagnetic semiconductor

$$E = E_0 \exp[j(wt - k_0 \underline{n}x)]. \qquad (2.8)$$

The vacuum wave number k_0 is given by

$$k_0 = 2\pi/\lambda_0, \qquad (2.9)$$

with λ_0 as the vacuum wavelength.

The physical meaning of n and κ will be clearer after separating the imaginary and real parts in the wave equation, Eq. 2.8.

$$E = E_0 \exp[j(wt - k_0 \underline{n}x)]\exp(-k_0\kappa x) \qquad (2.10)$$

The refractive index n describes the wave vector in a material

$$k = k_0 \cdot n \text{ or } \lambda = \lambda_0/n, \qquad (2.11)$$

whereas the absorption index describes the attenuation of the field strength $\overline{E}$ on a $k_0 x$ scale.

For measurement purposes the absorption coefficient α is more convenient, which is a measure of the intensity attenuation on a pure length scale. Because the intensity is proportional to the square of the $\overline{E}$ field, this absorption coefficient reads as follows:

$$\alpha = 2k_0\kappa \qquad (2.12)$$

The relations between the optical constants n and κ and the dielectric function are summarized in the following equation:

$$\varepsilon_r' = n^2 - \kappa^2 = n^2 - (\alpha^2/4k_0^2) \qquad (2.13)$$

$$\varepsilon_r'' = 2n\kappa = n\alpha/k_0 \qquad (2.14)$$

The frequency dependencies of the dielectric function components ε_r' and ε_r'' are not independent; they are linked by the Kramers–Kronig relationship

$$\varepsilon_r'(\omega^*) - 1 = \frac{2}{\pi} \int_0^\infty \frac{\omega \varepsilon_r''(\omega)}{\omega^2 - (\omega^*)^2} \, d\omega \qquad (2.15)$$

and

$$\varepsilon_r''(\omega^*) = -\frac{2\omega^*}{\pi} \int_0^\infty \frac{\varepsilon_r'(\omega) - 1}{\omega^2 - (\omega^*)^2} \, d\omega . \qquad (2.16)$$

That means if the whole frequency dependence of one component is known the other can be calculated by a numerical integration.

2.3 Absorption Processes

In the visible and near-infrared spectral regime the fundamental absorption processes [4] can be easily explained considering the electronic band structure of semiconductors and keeping in mind the conservation laws of energy and momentum. The energy E_{ph} of a photon is rather high in this spectral range,

$$E_{ph}(\omega) = \hbar\omega = 1.24 eV / \lambda_0 \text{ (μm)}, \qquad (2.17)$$

with a rather small momentum $\vec{p}_{ph}$

$$\vec{p}_{ph}(\omega) = \hbar k_0 = \hbar\omega / c \text{ (vacuum)} \qquad (2.18)$$

compared to the momentum space $\vec{p}_{max}$ given by the first Brillouin zone

$$\vec{p}_{max} = \hbar k_{max} = \hbar\pi / a_0 \text{ (diamond lattice)} \qquad (2.19)$$

as $\vec{p}_{ph}/\vec{p}_{max} = 2a_0/\lambda_0$ is typically on the order of 10^{-3} (Si, $\lambda_0 = 1.1$ μm). The phonon energies are rather low (Si: below 65 meV), but they cover the whole momentum space. As an example Fig. 2.3 shows the energy versus wave vector diagram of transversal optical (TO) and transversal acoustic (TA) phonons in Si.

TO phonons have got energies rather insensitive to the wave number. TA phonon energies start roughly linear with wave numbers.

Interband absorption (valence band to conduction band) requires as minimum energy the bandgap transition E_g. This gives a clear cutoff wavelength above which no interband absorption is possible.

$$\lambda_{cutoff} = 1.24 \ \mu m / E_g \ (eV) \tag{2.20}$$

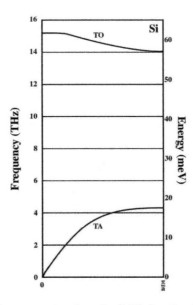

Figure 2.3 The phonon spectra along the (100) direction in Si. Shown are the transverse modes TO and TA, respectively (O: optical, A: acoustic).

Absorption above E_g is strongly dependent on the type of transitions (direct or indirect) because indirect transitions need phonons to satisfy momentum conservation. One obtains easy relations for the absorption coefficient, assuming effective mass approximation for the electronic band structure.

Direct transition: $\quad \alpha \cdot \omega = A \ (E_{ph} - E_g)^{1/2} \tag{2.21}$

Within a limited energy range near the band extremum (where effective mass approximation is valid for steady functions) this relation is often simplified as α^2 proportional to $(E_{ph} - E_g)$.

Indirect transition: The onset of absorption caused by indirect transitions is much weaker due to the phonon contribution for momentum conservation. It is described by

$$\alpha = A' \ (E_{ph} - E_g)^2 \tag{2.22}$$

The extracted bandgap E_g is in reality a temperature-dependent mixture of two curves for $E_g + E_{phon}$ and $E_g - E_{phon}$ because absorption and emission of a phonon are possible. As an example silicon absorption [5] is shown (Fig. 2.4) because Si is completely indirect in its band structure up to 3.2 eV. The square root of the absorption coefficient versus the energy scale should give a straight line for indirect semiconductors. This is fulfilled near the band edge when the effective mass model is a good description of the density of states of the bands. Photocurrent spectroscopy may be used for absorption measurements of thin layers of materials [6] in order to assess direct or indirect fundamental absorption and to estimate the band edge energy. However, near energetic distances between indirect and direct bandgap as given in pure Ge or GeSn alloys result in higher resonant indirect optical absorption [7].

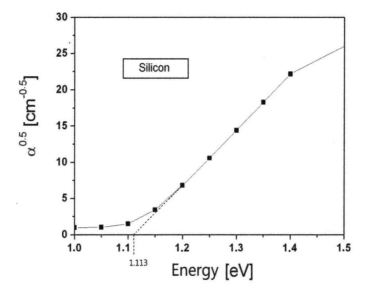

Figure 2.4 The room temperature absorption α of Si, depicted as $\alpha^{0.5}$ versus E_{ph}.

2.4 Direct Group IV Semiconductors

Unstrained alloys change their lattice cell volume V with composition. The relative volume change $\Delta V/V$ with respect to a reference cell depends on the lattice mismatch f between the alloy and the reference. We treat the corresponding bandgap changes within a linear approximation, which delivers enough accuracy within a few percent volume change because the data for deformation potentials and indirect/direct conduction band crossover are uncertain by at least 10%.

$$\Delta V/V = 3f \qquad (2.23)$$

The lattice mismatch f is 0.043*x for SiGe, with Si as reference, and 0.147*y for GeSn, with Ge as reference, respectively. The molar concentration of Ge in SiGe is given by x, and the molar concentration of Sn in GeSn is given by y.

With increasing lattice cell volume of GeSn, the energy difference ΔE_c between the direct conduction band and the indirect one shrinks from 136 meV for Ge to 0 meV for GeSn at the crossover concentration. The crossover concentration of Sn in GeSn is assessed to be between 7% and 11% [8]. The uncertainty stems from the contribution of residual strain and the low detectivity of weak indirect transitions near the strong direct transitions. The following numerical values for the linear approximation are consistent with data from Ref. [9]. In this work, the crossover is obtained either with 9% Sn in GeSn or 2% biaxial tensile strain. A linear superposition delivers for the energy difference ΔE_c of a biaxial strained GeSn:

$$\Delta E_c = 0.136 \text{ eV} - 1.51y - 6.7 \text{ eps} \qquad (2.24)$$

Crossover is obtained if Sn content y and biaxial strain eps fulfil the relation

$$1.51y + 6.7 \text{ eps} = 0.136 \text{ eV} \qquad (2.25)$$

Applying strain changes the lattice cell volume. An increase is obtained with a tensile strain. The volume increase is much larger for biaxial strain than for uniaxial strain. The cubic cell is distorted to a tetragonal cell for a (100) substrate surface. The following relations for uniaxial and biaxial strains then give the cell volume changes.

Uniaxial strain: $\quad \Delta V/V = \text{eps}*(1-2\,\nu) = \text{eps}*0.46 \qquad (2.26)$

Biaxial strain: $\quad \Delta V/V = 2\text{eps}*[(1-2\,\nu)/(1-\nu)] = \text{eps}*1.26 \qquad (2.27)$

The Poisson number ν describes the contraction in a perpendicular direction to the tensile elastic stress; its value is 0.27 for Ge.

Comparison of Eqs. 2.26 and 2.27 proves that the volume change with biaxial strain is nearly three times higher than that with uniaxial strain. Indeed, crossover with uniaxial strain requires more than 4% tension, in agreement with the simple volume considerations.

Both tensile biaxial strain and Sn alloying reduce the bandgap to about 0.5 eV for crossover.

A rough approximation for the direct bandgap E_{gdir} delivers as first overview

$$E_{gdir} = 0.8 \text{ eV} - 3.3\,y - 14.7 \text{ eps} \tag{2.28}$$

This approximation is only valid for biaxial tensile strain. Strain splits the degenerate values of heavy holes and light holes. The lowest energy transition is related to the light hole valence band for biaxial tensile strain, whereas compressive biaxial strain moves up the heavy hole band. This causes a fundamental asymmetry of bandgaps concerning the sign of the strain (remark: tensile strain is counted positive, compressive strain negative).

References

1. Landolt-Börnstein (2001). *Group III-Condensed Matter*, Vol. 41, Semiconductors, electronically available at http://materials.springer.com.
2. C. F. Klingshirn (2012). *Semiconductor Optics*, 4th Ed. (Springer).
3. D. E. Aspnes and A. A. Studna (1983). *Phys. Rev. B*, **27**, 985.
4. E. Kasper and D. J. Paul (2005). *Silicon Quantum Integrated Circuits* (Springer).
5. J. Humlicek (2000). *Properties of Silicon Germanium and SiGe: Carbon*, *EMIS Datareviews Series*, Vol. 24 (INSPEC, IEE, London), p. 249.
6. M. Oehme, M. Kaschel, J. Werner, O. Kirfel, M. Schmid, B. Bahouchi, E. Kasper and J. Schulze (2010). Germanium on silicon photodetectors with broad spectral range, *J. Electrochem. Soc.*, **157**, 144–148.
7. J. Menendez, M. Noel, J. C. Zwickels and D. Lockwood (2017). Resonant indirect optical absorption in germanium, *Phys. Rev. B*, **96**, 121201.
8. J. Liu (2014). Monolithically integrated Ge-on-Si active photonics, *Photonics*, **1**, 162–197.
9. X. Wang and J. Liu (2018). Emerging technologies in Si active photonics, *J. Semicond.*, **39**, 061001, doi: 10.1088/1674-4926/39/6/061001.

Chapter 3

Planar Waveguides

Optical signal transmission via fiberglass waveguides revolutionized telecommunication over long distances. The wavelength regimes around 1.3 μm and 1.55 μm are chosen because of extremely low absorption and dispersion windows for the silica fiberglass used. The loss requirements for planar waveguides on substrates with wafer-size (about 30 cm) to chip-size (about 1 cm) dimensions are less stringent. The wavelength window is open in the transparency regime, which is the near-infrared regime above 1.2 μm for silicon as the waveguide material. The application scenario and the availability of active devices define the upper limit. Now, germanium (Ge) on silicon (Si) active devices cover the wavelength range up to 1.55 μm. However, ongoing research with new semiconductor materials (strained Ge, GeSn, and low-bandgap III/V compounds) will extend the available range into the mid-infrared beyond 2.5 μm. Silicon, as the main material in the microelectronics industry, has attained great success; thus one wishes to utilize silicon as the base for photonic systems, too. The planar waveguide is an essential building block for photonic systems. So far, the silicon-on-insulator (SOI) platform has been the most popular structure for silicon waveguides. It has many advantages, such as high speed, low loss, small size, optoelectronic integration, and compatibility with the mature complementary metal-oxide semiconductor technology.

Silicon-Based Photonics
Erich Kasper and Jinzhong Yu
Copyright © 2020 Jenny Stanford Publishing Pte. Ltd.
ISBN 978-981-4303-24-8 (Hardcover), 978-981-4303-25-5 (eBook)
www.jennystanford.com

22 | *Planar Waveguides*

Electromagnetic theory of light is the basis for understanding guided wave optics. Research on the effects of light in the waveguide, such as propagation, scattering, polarization, and diffraction, becomes the theoretical basis for a variety of waveguide devices. For silicon material, we want to study transmission, coupling, and interaction with the external field of light, especially the single-mode condition and transmission characteristic of the SOI waveguide. In the first part of the chapter, we will focus on the mode characteristics and single-mode conditions for the SOI waveguides, including the slab, strip, and rib waveguides, and then continue on the loss and polarization dependence in the SOI waveguide.

3.1 Modes in the Slab Waveguide

As a kind of electromagnetic wave, the propagation of light in a waveguide must satisfy Maxwell's equations. The material properties, structure, and dimensions of an optical waveguide are the boundary conditions for solving Maxwell's equations. To solve Maxwell's equations in the optical waveguide, we will get multiple different sets of eigenvalues and their corresponding eigenfunctions. The eigenfunction is the corresponding field distribution of the electromagnetic field components in the waveguide cross section, that is, it corresponds to the discrete mode of optical wave propagation. The eigenvalue is the propagation constant β of the corresponding mode in a given waveguide.

By solving Maxwell's equations, we first discuss the field distributions of various modes in the slab waveguide and then we extend them to the field distributions in strip and rib waveguides.

Maxwell's equations can be specifically expressed in the following forms (Eqs. 3.1–3.4):

$$\nabla \times \vec{E}(r,t) = -\frac{\partial \vec{B}(r,t)}{\partial t} \tag{3.1}$$

$$\nabla \times \vec{H}(r,t) = \vec{j} + \frac{\partial \vec{D}(r,t)}{\partial t} \tag{3.2}$$

$$\nabla \vec{D}(r,t) = \rho \tag{3.3}$$

$$\nabla \vec{B}(r,t) = 0 \tag{3.4}$$

Here, $\vec{j}$ is the vector of free current density, ρ is the volume density of free charge, ∇ is the Hamiltonian operator, $\vec{E}$ is the electric field vector, $\vec{H}$ is the magnetic field vector, $\vec{D}$ is the electric flux density or electric displacement vector, and $\vec{B}$ is the magnetic flux density vector. They are all related to the time variation t and the position vector r. For a passive dielectric medium, it has no current and charge sources, so $\rho = 0$ and $\vec{j} = 0$, and Maxwell's equations can be simplified.

In the classical theory, there are the constitutive relations between the flux densities $\vec{D}$ and $\vec{B}$ as well as the fields $\vec{E}$ and $\vec{H}$. For a linear and isotropic medium, the relations are given by

$$\vec{D} = \varepsilon \vec{E} \tag{3.5}$$

$$\vec{B} = \mu \vec{H} \tag{3.6}$$

where ε is the dielectric permittivity of the medium and μ is the magnetic permeability of the medium. For a linear dielectric medium, the permittivity ε and permeability μ are independent of field intensities, but most dielectric mediums become nonlinear when the electric field intensity is relatively high. For a lossless medium, ε and μ are real scalar; while for an absorbing medium, they are complex scalar. If we substitute the constitutive relations (Eqs. 3.5 and 3.6) into Maxwell's equations, assuming that the medium is homogeneous, we can derive the following basic wave equations for $\vec{E}$ and $\vec{H}$:

$$\nabla^2 \vec{E} - \mu\varepsilon \frac{\partial^2 \vec{E}}{\partial t^2} = 0 \tag{3.7}$$

$$\nabla^2 \vec{H} - \mu\varepsilon \frac{\partial^2 \vec{H}}{\partial t^2} = 0 \tag{3.8}$$

Generally, electromagnetic waves are related to time by a sinusoidal relationship. The electromagnetic wave is radiated with a single frequency or can even be decomposed into many single-frequency waves by using Fourier transforms. Hence, the electric field vector and magnetic field vector can be written in the following forms:

$$\vec{E}(r,t) = \vec{E}(r)\exp(-i\omega t) \tag{3.9}$$

$$\vec{H}(r,t) = \vec{H}(r)\exp(-i\omega t) \tag{3.10}$$

Here $\vec{E}(r)$ and $\vec{H}(r)$ are the complex amplitude vectors and ω is the angular frequency. Substituting Eqs. 3.9 and 3.10 into Maxwell's equations (Eqs. 3.1 and 3.2), we get

$$\nabla \times \vec{E}(r) = i\omega\mu_0 \vec{H}(r) \tag{3.11}$$

and

$$\nabla \times \vec{H}(r) = -i\omega\varepsilon_0 n^2 \vec{E}(r), \tag{3.12}$$

where ε_0 and μ_0 are free space permittivity and free space permeability, respectively. For nonmagnetic dielectric mediums, $\mu = \mu_0$ and n is the refractive index of the medium and satisfies the relation $\varepsilon = \varepsilon_0 n^2$.

Now we solve Maxwell's equations in the planar dielectric waveguide, and then we can further analyze the electromagnetic wave modes in the waveguide. The slab waveguide is the simplest optical waveguide as shown in Fig. 3.1, and there are accurate analytical solutions to this waveguide. The slab waveguide consists of a guiding layer (or core layer), a substrate layer, and a cover layer or cladding layer. Their refractive indexes are n_1, n_2, and n_3, respectively, and commonly $n_1 > n_2 \geq n_3$. In the slab waveguide, the thickness of the core layer is d, which is much smaller than the waveguide width. So this waveguide can be considered infinite in the horizontal direction (y axis), namely

$$\frac{\partial \vec{E}}{\partial y} = \frac{\partial \vec{H}}{\partial y} = 0. \tag{3.13}$$

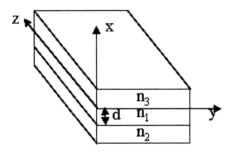

Figure 3.1 Planar dielectric waveguide.

Going by the coordinate axes shown in Fig. 3.1, the electric and magnetic fields are not functions of y for an electromagnetic wave propagating along the z axis. We assume the electric field polarizing along the y axis; the solutions to the wave equations should have the following forms:

$$E_y(x,z,t) = E(x)\exp[i(\beta z - \omega t)] \tag{3.14}$$

$$H_x(x,z,t) = H(x)\exp[i(\beta z - \omega t)] \tag{3.15}$$

If we substitute Eqs. 3.14 and 3.15 into Eqs. 3.11 and 3.12, we get the following:

$$\left.\begin{array}{l} \beta E_y = -\omega\mu H_x \\[2mm] \dfrac{\partial E_y}{\partial x} = i\omega\mu H_z \\[2mm] i\beta H_x - \dfrac{\partial H_z}{\partial x} = -i\omega\varepsilon E_y \end{array}\right\} \tag{3.16}$$

$$\left.\begin{array}{l} \beta H_y = \omega\varepsilon E_x \\[2mm] \dfrac{\partial H_y}{\partial x} = -i\omega\varepsilon E_z \\[2mm] i\beta E_x - \dfrac{\partial E_z}{\partial x} = i\omega\mu H_y \end{array}\right\} \tag{3.17}$$

As discussed in classical theory, there are two possible electromagnetic field polarizations, that is, the transverse electric (TE) field and the transverse magnetic (TM) field. The electric field of a TE wave exists only in the transverse direction, which means that there is no electric field component in the propagation direction; the magnetic field of a TM wave exists only in the transverse direction, which means that there is no magnetic field component in the propagation direction. Waves in a slab waveguide can be also classified as TE and TM waves. Looking at the above two sets of equations, these two sets of equations are independent of each other, and their solutions are also independent. There are only electromagnetic components E_y, H_x, and H_z in the Eqs. 3.16 and only electromagnetic components H_y, E_x, and E_z in Eqs. 3.17. Applying some assumptions, the solutions of these two sets of equations are a TE wave and a TM wave, respectively. If we consider one of the two waves, the TE wave, $E_z = 0$. Inserting it into Eqs. 3.17, we get

$$\partial H_y/\partial x = 0$$

that is, H_y is a constant independent of x. We assume this constant is zero because it doesn't have influence on the calculation results. From Eqs. 3.16, we know that E_x is zero too. Hence, the TE wave only has the electric field E_y component and the magnetic field H_x and H_z components. Similarly, inserting $H_z = 0$ into Eqs. 3.16, we can also obtain the TM wave with only the magnetic field H_y component and the electric field E_x and E_z components.

For simplification, Eqs. 3.16 and 3.17 can be reduced to the Helmholtz scalar equations of TE wave and TM wave, which are also called wave equations:

$$\frac{\partial^2 E_y}{\partial x^2} + (k_0^2 n_j^2 - \beta^2) E_y = 0 \tag{3.18}$$

$$\frac{\partial^2 H_y}{\partial x^2} + (k_0^2 n_j^2 - \beta^2) H_y = 0 \tag{3.19}$$

where $$k_0 = \omega \sqrt{\varepsilon_0 \mu_0} = 2\pi / \lambda \, .$$

The above equations are second-order differential equations, and they are solved by imposing the additional boundary conditions. The forms of the solutions to these second-order differential equations are related to the comparison of the magnitude between $k_0 n_j$ and β. If $k_0 n_j < \beta$, the solution is an oscillatory function (sinusoidal form); on the other hand, if $k_0 n_j > \beta$, the solution is an exponentially decaying function. On using the different propagation constant β to solve Eqs. 3.18 and 3.19, we can obtain different solutions with different field distribution in the slab waveguide. Each possible solution of β is called a mode. The modes in the slab waveguide can be classified as guided mode ($k_0 n_2 < \beta < k_0 n_1$), substrate mode ($k_0 n_3 < \beta \leq k_0 n_2$), and radiation mode ($0 \leq \beta \leq k_0 n_3$). For the guided mode, $k_0 n_2 < \beta < k_0 n_1$, solutions to the wave equation are exponentially decaying in the substrate layer and cladding layer; however, oscillatory waves form in the core layer. For the substrate mode, $k_0 n_3 < \beta \leq k_0 n_2$, solutions to the wave equation are exponentially decaying in the cladding layer, while oscillatory waves form in the substrate layer and the core layer. For the radiation mode, $0 \leq \beta \leq k_0 n_3$, there are oscillatory waves in all three layers of the waveguide. From the above discussion, we find that only when propagation constant β is in the range for the guided mode ($k_0 n_2 < \beta < k_0 n_1$), the electromagnetic wave can propagate in the guided layer. However, for the substrate mode, light waves will penetrate out of the waveguide through the substrate layer; for the radiation mode, light waves will penetrate out of the waveguide through the substrate layer and the cladding layer. Generally, it is deemed that the cladding and substrate layers have absorbing and scattering function for light waves, so light waves propagate in the two layers with a large loss and may even disappear. Therefore, we should avoid the latter two modes in the waveguide.

For simplicity, we take the TE wave as an example. To solve Eq. 3.18, according to the solution types in the guided mode case analyzed above, we can assume the electric field in the three layers of the slab waveguide to be written as follows:

$$E_y(x) = \begin{cases} A_1 \exp(-\delta x) & 0 \le x < +\infty \\ A\cos(\kappa x) + B\sin(\kappa x) & -d \le x \le 0 \\ A_2 \exp(\gamma x) & -\infty < x \le -d \end{cases}, \quad (3.20)$$

where A, B, A_1, and A_2 are amplitude coefficients to be determined by the boundary conditions, κ, γ, and δ, which are defined as follows:

$$\kappa = (k_0^2 n_1^2 - \beta^2)^{1/2}$$
$$\gamma = (\beta^2 - k_0^2 n_2^2)^{1/2} \quad (3.21)$$
$$\delta = (\beta^2 - k_0^2 n_3^2)^{1/2}$$

Considering the continuity of the electric field at $x = 0$ and $x = -d$, the boundary conditions are as follows:

$$\begin{cases} E_y(0^-) = E_y(0^+) \\ E_y(-d^-) = E_y(-d^+) \end{cases} \quad (3.22)$$

Combining Eq. 3.16 and boundary condition 3.22, we can get

$$\tan(\kappa d) = \frac{\kappa(\gamma + \delta)}{\kappa^2 - \gamma\delta} \quad (3.23)$$

Equation 3.23 is the characteristic equation for the TE modes of the slab waveguide. All the parameters (κ, γ, and δ) in the equation depend on the propagation constant β, so it is also the eigenvalue equation of β for TE modes of the slab waveguide. Because the equation is transcendental, the solutions for β to this equation need to be calculated numerically. We can consider κd as a variable. Using the mapping approach, we can get multiple sets of eigenvalue

Planar Waveguides

propagation constants and their corresponding field distributions, where each set of solutions corresponds to a propagation mode in the waveguide. Similarly, we can derive the characteristic equation for the TM modes of the slab waveguide:

$$\tan(\kappa d) = \frac{n_1^2 \kappa (n_2^2 \delta + n_3^2 \gamma)}{(n_2^2 n_3^2 \kappa^2 - n_1^4 \delta \gamma)} \tag{3.24}$$

Let us look at the solutions, Eqs. 3.20, in the slab waveguide again. The solution is sinusoidal in the core layer, which is a periodic function. Therefore, the possible value of β is discrete, that is, there are a limited number of modes that can exist in the slab waveguide.

From Eqs. 3.20 and 3.21 and the continuity of the field components in the x direction, comprehensively considering TE and TM modes, we can get

$$(n_1^2 - N^2)^{1/2} k_0 d = m\pi + \arctan\left(\sqrt{\frac{N^2 - n_2^2}{n_1^2 - N^2}}\eta_2\right) + \arctan\left(\sqrt{\frac{N^2 - n_3^2}{n_1^2 - N^2}}\eta_3\right),$$

$$\tag{3.25}$$

where $N = \beta/k_0$ is the effective refractive index of the mode; for the TE mode, $\eta_{2,3} = 1$; for the TM mode, $\eta_{2,3} = (n_1/n_{2,3})^2$. From the process of derivation, we find that Eq. 3.25 is equivalent to the characteristic Eqs. 3.23 and 3.24 for the TE or TM modes of the slab waveguide. m is called the mode number, $m = 0, 1, 2, \ldots$, where each m corresponds to an effective refractive N, which also proves the above analysis that the possible value of β is discrete.

Considering the most popular silicon waveguide system, that is, the silicon-on-insulator (SOI) platform, the three layers of the slab waveguide, as shown in Fig. 3.1, are air, silicon, and silica, with the corresponding refractive indices n_3, n_1, and n_2. The thickness of the middle silicon guiding layer is d. From the above discussion, solving the Helmholtz equations, we can obtain the analytical solutions of the waveguide modes. Figure 3.2 shows the field distributions of a few low-order TE and TM modes when the thickness $d = 1$ µm. Figure 3.2 shows that the field distributions penetrate deeper into the cladding and substrate layers when the mode number m increases. From Eq. 3.25, we find that when the mode number m increases, the effective refractive index N decreases, that is, the propagation constant β decreases. When the waveguide thickness d increases, the propagation constant β increases, and when the thickness is greater, the waveguide can support more guided modes. Figure 3.3 shows the relationship between the propagation constant of the first eight guided modes (including both TE and TM modes) and the silicon layer thickness. Due to the guided mode condition ($k_0 n_2 < \beta < k_0 n_1$), the mode number m cannot be infinitely large. Only a finite number of modes will be guided in the waveguide. If the waveguide thickness reduces to below some value, the waveguide will support only one mode with a polarization (TE or TM), which is the so-called single-mode waveguide. To reduce the interaction between the high-order modes, we usually prefer single-mode waveguides.

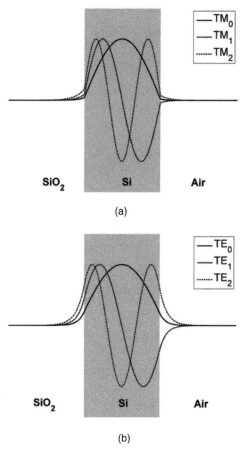

Figure 3.2 Field distributions for (a) TE_0, TE_1, and TE_2 modes and (b) TM_0, TM_1, TM_2 modes, when the silicon thickness d = 1 μm (1.55 μm wavelength).

30 | *Planar Waveguides*

If we take into consideration the guided mode condition $k_0 n_2 < \beta < k_0 n_1$, then $\beta = k_0 n_2$ is the cutoff condition of the slab waveguide guided modes. Therefore, we have the cutoff equation of guided modes

$$(n_1^2 - n_2^2)^{\frac{1}{2}} k_0 d = m\pi + \arctan\left(\sqrt{\frac{n_2^2 - n_3^2}{n_1^2 - n_2^2}}\, \eta_3\right). \tag{3.26}$$

From Eq. 3.26, we can also derive the condition of the slab waveguide maintaining single-mode operation as follows:

$$\frac{1}{k_0 \sqrt{n_1^2 - n_2^2}} \arctan\left(\sqrt{\frac{n_2^2 - n_3^2}{n_1^2 - n_2^2}}\, \eta_3\right) < d < \frac{1}{k_0 \sqrt{n_1^2 - n_2^2}}$$
$$\left[\pi + \arctan\left(\sqrt{\frac{n_2^2 - n_3^2}{n_1^2 - n_2^2}}\, \eta_3\right)\right], \tag{3.27}$$

where $k_0 = \dfrac{2\pi}{\lambda}$ is the wave number of light waves in vacuum and λ is the wavelength. The left part of the above equation is the cutoff thickness for TE_0 and TM_0 modes, and the right part is the cutoff thickness for TE_1 and TM_1 modes. Substituting the real refractive index of air, silicon, and silica into Eq. 3.27, we can get the single-mode condition 26.27 nm $< d <$ 274.58 nm for TE mode and 105.28 nm $< d <$ 353.59 nm for TM mode (1.55 μm wavelength). Obviously, in order to achieve single-mode transmission in the air/silicon/silica three-layer dielectric slab waveguide, we must reduce its size to the order of submicrons.

The above discussion reveals the basic methods to analyze the modes in the waveguide. For the analysis of the multilayer slab waveguides, strip waveguide and rib waveguide, the differences in methods are because of the differences in boundary conditions. We can analyze them by considering that the effective index method (EIM) is equivalent to the multilayer slab waveguide. Solving the propagation constants of the modes and the corresponding field distributions is the fundamental approach to analyzing the modes in the waveguide. The related issues are treated in the classical literature [1]; so we will not review them here.

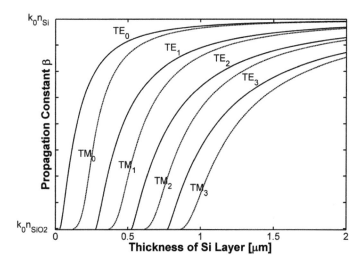

Figure 3.3 Relationship between the propagation constant β of the TE and TM modes and the silicon layer thickness (1.55 µm wavelength).

3.2 Strip Waveguides and Rib Waveguides

The slab waveguides confine light waves to only one dimension; the waveguide must be very thin to meet the single-mode condition. To confine light better, 2D confinement waveguides are required. There are two common basic SOI waveguide structures: strip waveguides (or photonic wires [2]) and rib waveguides, as shown in Fig. 3.4. The cross sections of the strip and rib waveguides can be rectangular (Fig. 3.4) or any other shape (such as a trapezoidal cross section). They confine light waves both in the horizontal and vertical directions, so an analytical solution of the modes in these waveguides cannot be directly obtained by mathematical derivation of the Helmholtz equations. When the widths of the strip and rib waveguides increase, the waveguide can support more guided modes, which is similar to the 1D slab waveguides. To analyze the modes in the waveguide, many approximate or numerical solutions have been developed, such as the EIM, the beam propagation method (BPM), the finite difference time domain, and the film mode–matching method.

To achieve single-mode operation, strip waveguide dimensions in both directions (height and width) should be below certain cutoff values. The SOI waveguide cutoff dimensions are usually smaller than 1 μm due to the high refractive index contrast between silicon (~3.5) and silica (~1.5). By using different numerical methods from the above description, one can accurately calculate the cutoff dimension of certain waveguide type and obtain the single-mode condition.

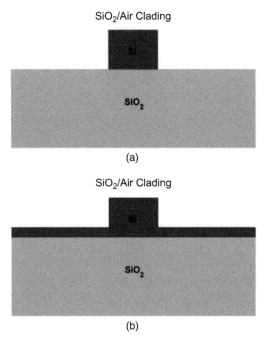

Figure 3.4 The cross-section structure diagram of (a) strip waveguides and (b) rib waveguides.

Figure 3.5 shows the single-mode condition for a strip waveguide with oxide cladding [3] when a full-vector finite difference method is used for calculation. The horizontal and vertical axes in Fig. 3.5 are the width w and height h of the core, respectively. The curves indicate the critical boundary under which the single-mode region lies. The size of the core for a single-mode Si strip waveguide is of the order of several hundred nanometers. Similar to 2D slab waveguides, the fundamental modes for TE and TM polarization in a 1D waveguide also have a cutoff condition. The single-mode region is located between the curves that are determined by the cutoff conditions for the fundamental and first-order modes. We have two curves for each TE and TM polarization in a width w versus height H presentation (Fig. 3.6), which define the single-mode region for both modes (black color) and adjacent regions in which either TE or TM is single mode.

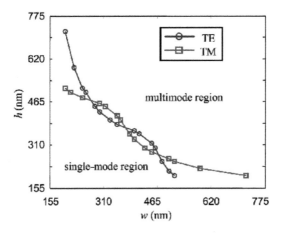

Figure 3.5 Single-mode conditions for the TE and TM modes of a strip waveguide with oxide cladding (1.55 μm wavelength). Reprinted with permission from Ref. [3] © The Optical Society.

Soon Lim et al. solved the cross-section field and effective refractive index of the strip waveguide by using a 3D imaginary BPM and analyzed the single-mode conditions [4] of the strip waveguide with oxide cladding at both operating wavelengths, 1310 nm and 1550 nm, in detail, as shown in Fig. 3.6. To satisfy single-mode conditions in both polarizations (TE and TM), two conditions have to be considered: (i) the cutoff condition of the first-order TE mode (upper limit) and (ii) the cutoff condition of the fundamental TM mode (lower limit). Let us consider a waveguide dimension of 300 nm × 350 nm; the result indicates that the waveguide is single mode at a wavelength of 1550 nm but not at a wavelength of 1310 nm; this implies that the single-mode condition is more relaxed at longer wavelengths and more stringent at shorter wavelengths. If the boundaries of the TM_0 and TE_1 cutoffs are fitted, we can obtain an experiential equation that describes the single-mode condition at wavelength 1550 nm as follows:

$$0.2 + 162 e^{-H/0.03} \leq W \leq 0.3 + 5.9 e^{-H/0.08} \tag{3.28}$$

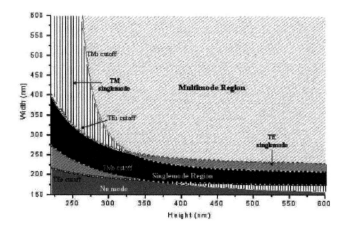

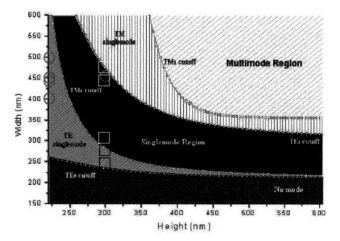

Figure 3.6 The single-mode condition for photonic wire at operation wavelengths of (a) 1330 nm and (b) 1550 nm. Reprinted with permission from Ref. [4] © The Optical Society.

For rib waveguides, in the middle of the cross section, there is a raised ridge region as the guided region that is similar to strip waveguides. But on both sides of the ridge the film is not completely removed, which is different to strip waveguides. The cross section of a rib waveguide in SOI is shown in Fig. 3.7. Due to the existence of the outside slab region, rib waveguides have a more complex mode characteristic, where high-order modes (modes other than the fundamental mode) can leak out from the slab region. Therefore, large rib waveguides even with micron-sized cross-sectional dimensions can behave as single-mode waveguides. However, strip waveguides support multiple modes when the cross-section dimensions reach a few hundred nanometers. Large rib waveguides, whose dimensions are closer to those of optical single-mode fibers, can achieve low-loss coupling to optical fibers. Rib waveguides have stronger confinement for light waves than strip waveguides, and we can fabricate electrodes on their slab region easily. So, the rib structure has been widely used for photonic waveguides.

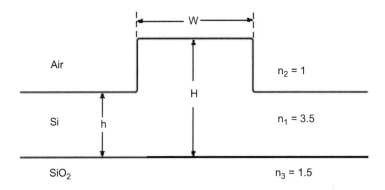

Figure 3.7 Cross section of a rib waveguide.

Petermann first proposed multimode rib waveguides with a large cross section [5]. Soref et al. used mode matching and BPMs to analyze the single-mode operation condition of optical GeSi-Si and Si-SiO$_2$ rib waveguides. They gave the following relation for single-mode conditions [6]:

$$\begin{cases} \dfrac{W}{H} \leq 0.3 + \dfrac{r}{\sqrt{1-r^2}} \\ r = \dfrac{h}{H} \geq 0.5 \end{cases}, \quad (3.29)$$

where W is the rib width, H is the overall rib height, and h is the slab height, as shown in Fig. 3.7; r is the ratio of slab height to rib height. The second equation in Eqs. 3.29 represents a rib waveguide that is shallow-etched to ensure single-mode characteristics in the vertical direction of the rib waveguide. If $r > 0.5$, the effective index (propagation constant) of high-order modes in the vertical direction of the ridge region is smaller than that of the fundamental mode in the slab region, which makes high-order modes leak out of the guided region, leaving only fundamental modes in the vertical direction of the ridge region. After r is determined, the ratio W/H is determined by the first equation in Eqs. 3.29, which ensures single-mode characteristics in the horizontal direction of the rib waveguide.

Later, Rickman et al. and Schmidchen et al. studied the single- and multimode conditions of SOI rib waveguides by using experimental methods [7, 8]. They found a considerable difference between Soref's formula and the experimental results. Soref's formula is more relaxed than experimental data. On the basis of the above differences, many researchers analyzed the single-mode condition of large rib waveguides further using different numerical methods. Pogossian et al. studied the single-mode condition of the rib waveguide by using the EIM and obtained the single-mode condition [9], which is in better agreement with Rickman's experimental data, as shown in Fig. 3.8. The single-mode condition can be written as:

$$\begin{cases} \dfrac{W}{H} \leq \dfrac{r}{\sqrt{1-r^2}} \\ r = \dfrac{h}{H} \geq 0.5 \end{cases}, \qquad (3.30)$$

where the variable definitions are the same as in Eq. 3.29.

Moreover, Xia et al. [10] analyzed the mode characteristics for rib waveguides with a trapezoidal cross section by using the BPM and then they obtained the single-mode condition for rib waveguides that is similar to Soref's formula.

The difference between numerical modeling and experiment may partly be explained by the different length scales. Theory considers stable modes in an infinite long waveguide, but in experiment light is coupled in and out after finite lengths.

When light waves couple with the waveguide from optical fibers, multiple different modes are exited in the waveguide due to discontinuous medium interfaces. Some modes constitute the guided modes in the waveguide, while the modes that cannot be guided in the waveguide leak out of the waveguide after a certain length [11]. The final stable field distribution in the waveguide is the superposition of the guided modes. It is possible to detect the high-order modes if the waveguide length is not long enough in the experiment and then to consider the waveguide as a multimode waveguide.

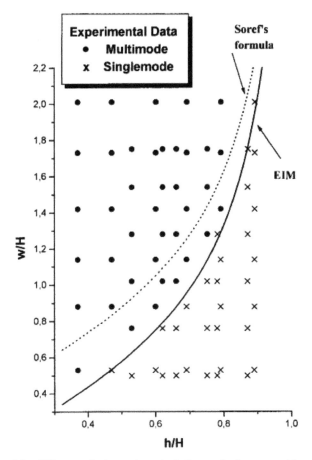

Figure 3.8 Different single-mode calculations of rib waveguides (Soref's formula, EIM) from Refs. [6, 9] compared to experimental data [7, 8] on a w/H versus h/H presentation.

However, the current trend is miniaturization and integration in silicon photonic devices to improve device performance and cost efficiency. When the cross-section dimensions of the rib waveguide reduce to around 1 μm or even to submicron levels, its single-mode condition is different to that of the large rib waveguide. The Yu [12] research group calculated the single-mode cutoff dimensions for quasi-TE and quasi-TM modes by using scalar and full-vector numerical simulation methods, respectively. It has been pointed out that small and deeply etched rib waveguides can satisfy the single-mode condition as long as the waveguide dimensions are designed reasonably. From the simulation data obtained, they give a fitting formula for the single-mode condition of small and deeply etched rib waveguides as follows:

$$
\begin{cases}
\dfrac{W}{H} \leq 0.05 + \dfrac{(0.94 + 0.25H)r}{\sqrt{1 - r^2}} \\
0.3 < r < 0.5, 1.0 \leq H \leq 1.5
\end{cases}
, \tag{3.31}
$$

where r is in the range of 0.3 to 0.5 because of polarization dependence.

Compared with the single-mode conditions given by Soref et al. and Pogossian et al., the simulation results of the single-mode condition for small-cross-section rib waveguides have the following characteristics.

- The ratio of slab height to rib height r is no longer limited to the condition $r > 0.5$. When the rib waveguide is deeply etched with $r < 0.5$, it can achieve single-mode transmission in the waveguide by choosing an appropriate waveguide width W.
- There is an obvious difference between the single-mode conditions for quasi-TE and quasi-TM modes, and the single-mode condition for the quasi-TM mode is more rigorous. Therefore, the single-mode condition for the quasi-TM mode should be the limiting condition for the waveguide design, because only when this is satisfied, both polarizations are single-mode ones.

3.3   Loss in a Silicon Optical Waveguide

With the development of the silicon micronanofabrication technology, single-mode propagation loss in a silicon waveguide has reduced to several decibels per centimeter from the initial several hundred decibels per centimeter. Low propagation losses of 0.8 dB/cm have been reported [13].

Loss reduction in the waveguide is very important for the quality of guided wave transmission. There are three main reasons for loss in the waveguide: absorption, scattering, and radiation. The losses caused by the three effects in a silicon waveguide are dependent on the waveguide design and micronanofabrication technology. In this section, we will discuss loss mechanisms in SOI waveguides.

The silicon in the SOI guided layer is transparent at the communication band in the range of 1.3 μm to 1.55 μm, so the loss of the SOI waveguide caused by intrinsic absorption is negligible. However, for some active photonic devices, free carriers are injected into the silicon by applying an external bias, which leads to free carrier absorption. The loss caused by free carrier absorption may be significant in the SOI waveguide. The absorption loss is proportional to the concentration of free carriers.

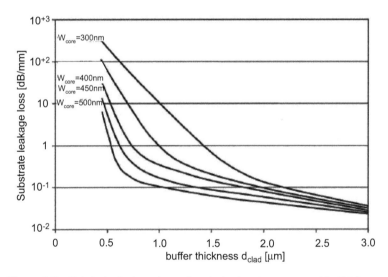

Figure 3.9 Substrate leakage loss of a photonic wire versus oxide thickness (parameter width *w*).

Substrate leakage is another important reason for propagation loss. In the case of submicron SOI waveguides, although they have strong confinement for light waves, a relatively large part of the mode field will leak into the cladding or substrate layer due to the small cross section. If the lower cladding layer of silica is too thin, the mode field even penetrates the buried oxide layer and couples with the silicon substrate, which leads to radiation loss. This leakage is larger due to smaller waveguide dimensions and a thinner buried oxide layer. Figure 3.9 shows the relationship between the loss of the TE mode in the waveguide with height $H = 0.22$ μm and the different widths caused by substrate leakage and the buried oxide thickness. It is clearly seen that the loss caused by substrate leakage decreases while the buried oxide thickness and waveguide width increase. Therefore, the buried oxide should be thick enough to reduce the substrate leakage loss.

For submicron waveguides, scattering caused by imperfections in the bulk waveguide material and roughness at the interface between different mediums in the waveguide is the main loss mechanism. The waveguide sidewalls are produced through lithography and etching process, so they are much rougher than the upper and lower interfaces of the silicon layer; and with the waveguide size shrinking further, the interaction between the mode field and rough sidewalls will be further enhanced, which leads to a sharp increase in the loss.

Sidewall roughness in the waveguide can change the waveguide width with random fluctuations, that is, the waveguide width is a random varying function along the propagation direction. Usually, we can use the standard deviation σ and the correlation length L_c to make a quantitative description for this function. These parameters can be measured through a variety of electron microscope experiments. Payne–Lacey theory indicates the scattering loss coefficient α (dB/cm) caused by the sidewall roughness in the waveguide as [14]:

$$\alpha = 4.34 \frac{\sigma^2}{k_0 \sqrt{2} w^4 n_{\text{clad}}} g(V) \cdot f_e(x, \gamma), \qquad (3.32)$$

where $g(V) = \dfrac{U^2 V^2}{1 + W}$ is a function depending only on the waveguide geometry with the normalized coefficients $U = k_0 d \sqrt{n_c^2 - n_{\text{eff}}^2}$, $V = k_0 d \sqrt{n_{\text{core}}^2 - n_{\text{clad}}^2}$, and $W = k_0 d \sqrt{n_{\text{eff}}^2 - n_{\text{clad}}^2}$. The function f_e (x, γ) is linked to the sidewall roughness

$$f_e(x, \gamma) = \frac{x \sqrt{1 - x^2 + \sqrt{(1 + x^2)^2 + 2x^2 \gamma^2}}}{\sqrt{(1 + x^2)^2 + 2x^2 \gamma^2}}$$

$$x = W \frac{L_c}{d}, \quad \gamma = \frac{n_{cl} V}{n_c W \sqrt{\Delta}}$$

$$\Delta = \frac{n_{core}^2 - n_{clad}^2}{2n_{core}^2}, \qquad (3.33)$$

where n_{core} and n_{clad} are the refractive indices of the core layer and the cladding layer, respectively; n_{eff} is the effective refractive index, w is the waveguide width, and k_0 is the free space wave number. Figure 3.10 shows the contour lines of the sidewall scattering loss versus the standard deviation σ and the correlation length L_c in the strip waveguide with a 200 nm × 200 nm cross-sectional dimension. The functions of $g(V)$ and $f_e(x, y)$ in Eq. 3.32 are slowly varying functions depending on the waveguide parameters, which have little effect on waveguide loss, while the waveguide width and the standard deviation are the main factors for the waveguide loss.

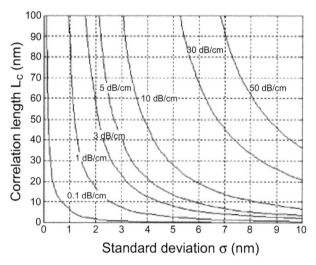

Figure 3.10 Contour lines of the sidewall scattering loss of a 200 nm × 200 nm strip waveguide as a function of (σ, L_c).

Measuring the loss in the waveguide is very important for designing photonic devices because the loss determines the quality of guided wave transmission in the waveguide. The propagation losses in the SOI single-mode strip waveguides have been measured by many research groups. There are many experimental methods associated with waveguide measurement, such as the cutback method, the Fabry–Perot resonance method, and the Fourier spectral analysis method. Table 3.1 shows some reported results.

Table 3.1 Comparison of the propagation loss measurement results for SOI single-mode strip waveguides (TE)

Reference	Height (nm)	Width (nm)	Propagation loss (dB/cm)	Wave length (nm)	Measurement method	Note
Sakai et al. [16]	320	400	25 ± 10	1550	Fabry–Perot	0.18 mm long
Lee et al. [13]	200	500	32 (traditional method) 0.8 (oxidation smoothing)	1540	Cutback	SiO$_2$ cladding
Almeida et al. [17]	270	470	5 ± 2	1550	Fabry–Perot	13 mm long
McNab et al. [18]	220	465	3.5 ± 2	1500	IR capture	4 mm long
Yamada et al. [19]	300	300	6	1550	Cutback	Oxidation smoothing, 16 mm long
Dumon et al. [14]	220	400	33.8 ± 1.7	1550	Fabry–Perot	1 mm long
		450	7.4 ± 0.9			
		500	2.4 ± 1.6			
Vlasov et al. [20]	220	445	3.6 ± 0.1	1500	Cutback	21 mm long

Bending of a waveguide leads to radiation losses in the bend. The magnitude of the bend loss strongly depends on the bend radius: the bend loss has a sharp increase with the radius curvature. It has been demonstrated in both theory and experiments that the bend loss in the submicron SOI strip waveguide is usually of the order of 0.1 dB, even with the bend radius $R = 1$ µm. Table 3.2 shows some bend loss measurement results of the single-mode SOI strip waveguides given in literature. To reduce the bend loss further, there are some methods: compact resonator structures are introduced into the bend to increase the transmissivity; the lateral offset is introduced into the junction between the straight waveguide and the bent waveguide to achieve better mode field matching.

Table 3.2 Comparison of bend loss measurement results for SOI single-mode strip waveguides (TE)

Reference	Height (nm)	Width (nm)	Radius (µm)	Loss (dB)	Wavelength (nm)
Lim [21]	200	500	1	0.5	1540
Sakai et al. [16]	320	400	1	1 ± 3	1550
Tsuchizawa et al. [22]	300	300	2	0.46	1550
			3	0.17	
Ahmad et al. [23]	340	400	Corner mirror	1	1550
Dumon et al. [24]	220	400	15	0.5	1550
			Corner mirror	1	
Vlasov et al. [20]	220	445	1	0.086 ± 0.005	1500
			2	0.013 ± 0.005	
			5	±0.005	

It is easier to obtain a small propagation loss in the rib waveguide because the side area is reduced. The propagation loss of a shallow-etched rib waveguide with ridge height $H = 0.34$ µm, width $W = 0.50$ µm, and slab height $h = 0.14$ µm is about 0.7 dB/cm [25]. However, the shallow-etched rib waveguide has weaker confinement in the lateral direction, so light in the ridge waveguide leaks out easily from the slab region. Usually, the ridge waveguide requires a longer bend radius to obtain a low bend loss. Therefore, the rib waveguides often use corner reflectors to achieve a 90º deflection of rays. The bend loss in the rib waveguide with corner reflectors was reduced to 0.32 ± 0.02 dB (92.9% bend efficiency) for TE polarization at λ = 1.55 µm [26].

3.4 Polarization Dependence of Silicon Waveguides

Polarization dependence in SOI optical waveguides of small dimensions is rather strong. As already shown in Fig. 3.2, there is a difference between the TE and TM field strength distributions in the waveguide. Figure 3.11 shows the field distributions and loss spectra for TE and TM modes in a single-mode SOI strip waveguide.

Usually, polarization dependence is described by waveguide birefringence, which is defined as the effective index or group refractive index difference between TE and TM modes,

$$\Delta n_{\text{eff}} = n_{\text{eff}}^{\text{TE}} - n_{\text{eff}}^{\text{TM}}.$$

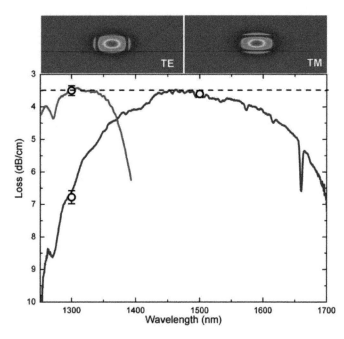

Figure 3.11 The mode field distributions (above) and loss spectra (below) for TE and TM modes in a single-mode SOI strip waveguide.

Figure 3.12 shows the relationship between the effect refractive index and the waveguide width, where four colored lines represent TE and TM modes in the strip and rib waveguides, respectively. The effective refractive index for the TE mode can be equal to the one for the TM mode in the waveguide at some specific dimensions in order to achieve zero birefringence. However, it should be noted that a deviation in the waveguide width will cause a change in the birefringence. When the strip waveguide width has a 10 nm deviation, the birefringence deviation reaches the order of 10^{-2}.

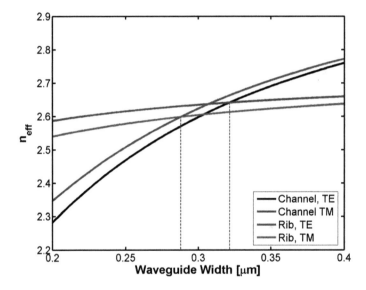

Figure 3.12 Effective refractive index of TE and TM modes in the strip and rib waveguides.

Birefringence control from the geometry and dimension of the waveguide requires precise dimension control with a high requirement for fabrication tolerance. The waveguide birefringence of a given waveguide may be changed by two mechanisms: asymmetry of the structure and the opto-elastic effect due to anisotropic stress in the waveguide core layer. Therefore, we could consider these two aspects to reduce polarization dependence.

A silica cladding usually covers the silicon core layer of SOI waveguides. The stress in the upper cladding layer causes anisotropic stress distribution in the core and at the edge of the silicon layer, and the refractive indices of silicon and silica are changed by the elastic-optic effect; a stress-related birefringence component is introduced. The birefringence caused by stress is related to the oxide density, oxide thickness, and waveguide geometric. Therefore, the waveguide birefringence can be effectively reduced by properly changing the upper cladding thickness and the stress in the material.

3.5   Summary

Silicon has shown great advantage for photonic waveguiding, especially for optoelectronic integration. SOI has become the most popular platform for silicon photonic waveguides due to the high refractive index contrast between the semiconductor and the surrounding silicon oxide. To discuss basic waveguide properties, we began with Maxwell's equations, analyzed the mode field distributions in the slab waveguide, extended that to the strip and rib waveguides, and then gave a comprehensive analysis of single-mode conditions for the waveguides. Basic loss mechanisms and polarization properties were discussed, with reference to experimental results and numerical calculations. Nanowire waveguides are the backbone of chip-sized photonic systems; they transport optical signals between sources and receivers, splitters divide the signal into various directions, and different signals are added at waveguide combiners. The high refractive index contrast of SOI waveguides allows dense packing of many waveguides, with strong bending on a silicon chip.

Acknowledgements

We thank Xuejun Xu and Kang Xiong for collection of basic informations and figure drawings.

References

1. M. Born and E. Wolf (1999). *Principles of Optics* (Cambridge Univ. Press, Cambridge, UK).
2. W. Bogaerts, D. Taillaert, B. Luyssaert, P. Dumon, J. Van Campenhout, P. Bienstman, D. Van Thourhout, R. Baets, V. Wiaux and S. Beckx (2004).

Basic structures for photonic integrated circuits in silicon-on-insulator, *Opt. Express*, **12**, 1583–1591.

3. D. Dai, Y. Shi and S. He (2006). Characteristic analysis of nano silicon rectangular waveguides for planar light-wave circuits of high integration, *Appl. Opt.*, **45**, 4941–4946.

4. S. T. Lim, C. E. Png, E. A. Ong and Y. L. Ang (2007). Single mode, polarization-independent submicron silicon waveguides based on geometrical adjustments, *Opt. Express*, **15**, 11061–11072.

5. K. Petermann (1976). Properties of optical rib-guides with large cross-section, *Archiv für Elektronik und Überetragungstechnik*, **30**, 139–140.

6. R. A. Soref, J. Schmidtchen and K. Petermann (1991). Large single-mode rib waveguides in GeSi-Si and Si-on-SiO$_2$, *IEEE J. Quantum Electron.*, **27**, 1971–1974.

7. A. G. Richman, G. T. Reed and F. Namavar (1994). Silicon-on-insulator optical rib waveguide loss and mode characteristics, *J. Lightwave Technol.*, **12**, 1771–1776.

8. J. Schmidtchen, A. Splett, B. Schuppert, K. Petermann and G. Burbach (1991). Low loss singlemode optical waveguides with large cross-section in silicon-on-insulator, *Electron. Lett.*, **27**, 1486.

9. S. P. Pogossian, L. Vescan and A. Vonsovici (1998). The single-mode condition for semiconductor rib waveguides with large cross section, *J. Lightwave Technol.*, **16**, 1851–1853.

10. J. Xia and J. Yu (2004). Single-mode condition for silicon rib waveguides with trapezoidal cross-section, *Opt. Commun.*, **230**, 253–257.

11. J. Lousteau, D. Furniss, A. B. Seddon, T. M. Benson, A. Vukovic and P. Sewell (2004). The single-mode condition for silicon-on-insulator optical rib waveguides with large cross section, *J. Lightwave Technol.*, **22**, 1923–1929.

12. X. Xu, S. Chen, J. Yu and X. Tu (2009). An investigation of the mode characteristics of SOI submicron rib waveguides using the film mode matching method, *J. Opt. A: Pure Appl. Opt.*, **11**, 015508.

13. K. K. Lee, D. R. Lim, L. C. Kimerling, J. Shin and F. Cerrina (2001). Fabrication of ultralow-loss Si/SiO2 waveguides by roughness reduction, *Optics Lett.*, **26**, 1888–1890.

14. P. Dumon, W. Bogaerts, V. Wiaux, J. Wouters, S. Beckx, J. Van Campenhout, D. Taillaert, B. Luyssaert, P. Bienstman, D. Van Thourhout and R. Baets (2004). Low-loss SOI photonic wires and ring resonators fabricated with deep UV lithography, *IEEE Photonics Technol. Lett.*, **16**, 1328–1330.

15. F. Grillot, L. Vivien, S. Laval, D. Pascal and E. Cassan (2004). Size influence on the propagation loss induced by sidewall roughness in ultrasmall SOI waveguides, *IEEE Photonics Technol. Lett.*, **16**, 1661–1663.

16. A. Sakai, G. Hara and T. Baba (2001). Propagation characteristics of ultrahigh-Δ optical waveguide on silicon-on-insulator substrate, *Jpn. J. Appl. Phys.*, **40**, L383–L385.

17. V. R. Almeida, R. R. Panepucci and M. Lipson (2003). Nanotaper for compact mode conversion, *Opt. Lett.*, **28**, 1302–1304.

18. S. J. McNab, et al. (2003). Ultra-low loss photonic integrated circuit with membrane-type photonic crystal waveguides, *Opt. Express*, **11**, 2927–2939.

19. K. Yamada, et al. (2004). Microphotonics devices based on silicon wire waveguiding system, *IEICE Trans. Electron.*, **E87-C**(3), 351–358.

20. Y. A. Vlasov, et al. (2004). Losses in single-mode silicon-on-insulator strip waveguides and bends, *Opt. Express*, **12**, 1622–1631.

21. D. R. Lim (2000). Device integration for silicon microphotonics platforms, PhD thesis, MIT.

22. T. Tsuchizawa, T. Watanabe, E. Tamechika, T. Shoji, K. Yamada, J. Takahashi, S. Uchiyama, S. Itabashi and H. Morita (2002). Fabrication and evaluation of submicron-square Si wire waveguides with spot-size converters, *Paper TuU2 presented at LEOS Annual Meeting*, Glasgow, UK, p. 287.

23. R. U. Ahmad, F. Pizzuto, G. S. Camarda, R. L. Espinola, H. Rao and R. M. Osgood, Jr. (2002). Ultracompact corner-mirrors and T-branches in silicon-on-insulator, *IEEE Photonics Technol. Lett.*, **14**, 65.

24. P. Dumon, W. Bogaerts, J. Van Campenhout, V. Wiaux, J. Wouters, S. Beckx and R. Baets (2003). Low-loss photonic wires and compact ring resonators in silicon-on-insulator, *LEOS Benelux Annual Symposium 2003*, Netherlands.

25. M. A. Webster, R. M. Pafchek, G. Sukumaran and T. L. Koch (2005). Low-loss quasi-planar ridge waveguides formed on thin silicon-on-insulator, *Appl. Phys. Lett.*, **87**, 231108-1-3.

26. Y. Qian, S. Kim, J. Song, G. P. Nordin and J. Jiang (2006). Compact and low loss silicon-on-insulator rib waveguide 90° bend, *Opt. Express*, **14**, 6020–6028.

Chapter 4

Microring Resonators

4.1 Introduction

Resonators have attracted considerable attention in the area of integrated photonics. Early in 1969, microring resonators were proposed by Marcatili at Bell Labs [1]. Due to the advanced fabrication techniques, integrated microring resonators have made great progress in the last few decades. Microring resonators have many merits, compared with other kinds of resonators. First, mirrors or gratings are not needed to get optical feedback; thus, microring resonators are not so hard to fabricate and are very suitable for on-chip optical integration. Second, the microring resonator is a traveling-wave resonator, and the optical paths for input, transmission, and reflection are separated, making it very compatible with other optical components. Third, multiple microring resonators can be coupled in series or parallel to achieve a box-like response and a large free-spectrum range (FSR), making it very promising in a dense wavelength division multiplexing system.

Silicon-on-insulator (SOI) is a promising material platform for optical integration, which is low cost and compatible with complementary metal-oxide-semiconductor (CMOS) processes and has a high index contrast to obtain compact optical devices.

Silicon-Based Photonics
Erich Kasper and Jinzhong Yu
Copyright © 2020 Jenny Stanford Publishing Pte. Ltd.
ISBN 978-981-4303-24-8 (Hardcover), 978-981-4303-25-5 (eBook)
www.jennystanford.com

Microring Resonators

Nowadays, as a basic component, the microring resonator has been widely employed to construct various optical devices, such as filters, modulators, switches, logic gates, delay lines, and sensors. For this chapter, the basic theory is illustrated in Section 4.1. Then, the optical properties and device designs are introduced in Sections 4.2 and 4.3, respectively. After that, the fabrication and measurement of the SOI-based microring resonator are presented in Section 4.4, followed by their applications in Section 4.5. Finally, we speculate about trends in silicon-based microring resonator development.

4.2 Principle of the Microring Resonator

4.2.1 Single Microring

A single microring [2–4] coupled to a bus waveguide is the simplest configuration, having an input port and an output port, as shown in Fig. 4.1a, which is equivalent to a Gires–Tournois resonator in Fig. 4.1b.

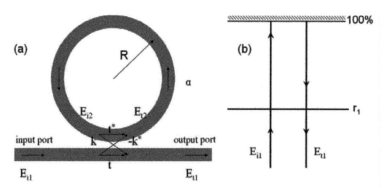

Figure 4.1 (a) Microring resonator coupled to a bus waveguide; (b) Gires–Tournois resonator.

If the waveguide has only a single mode, the coupling region is lossless, and polarization converting does not exist, we can treat all the optical loss in the ring with an attenuation factor. Then, the transfer matrix of the microring resonator can be described as

$$\begin{bmatrix} E_{t1} \\ E_{t2} \end{bmatrix} = \begin{bmatrix} t & k \\ -k^* & t^* \end{bmatrix} \begin{bmatrix} E_{i1} \\ E_{i2} \end{bmatrix} \qquad (4.1)$$

where E_{t1}, E_{t2}, E_{i1}, and E_{i2} denote the optical amplitude in the positions of microring as shown in Fig. 4.1a, k is the coupling coefficient between the microring and bus waveguide, and t is the transmitting coefficient. As the coupling region is assumed lossless, k and t satisfy the following relation:

$$|k^2| + |t^2| = 1 \tag{4.2}$$

After propagating in a single circle,

$$E_{i2} = \alpha e^{j\theta} E_{t2}, \tag{4.3}$$

where α is the attenuation factor in the microring, $\theta = n_{\text{eff}} L / \lambda_0$ is the phase shift in a circle, $L = 2\pi r$ is the ring perimeter, r is the ring radius, n_{eff} is the effective refractive index of the waveguide mode, and λ_0 is the optical wavelength in vacuum. According to Eq. 4.1 and Eq. 4.3, we obtain the following:

$$\frac{E_{t1}}{E_{i1}} = \frac{t - \alpha e^{j\theta}}{1 - \alpha t^* e^{j\theta}} \tag{4.4}$$

$$\frac{E_{i2}}{E_{i1}} = \frac{-\alpha k^* e^{j\theta}}{1 - \alpha t^* e^{j\theta}} \tag{4.5}$$

$$\frac{E_{t2}}{E_{i1}} = \frac{-k^*}{1 - \alpha t^* e^{j\theta}} \tag{4.6}$$

Figure 4.2a shows an add-drop filter based on a microring resonator coupled with two bus waveguides, which has four ports: input port, throughput port, add port, and drop port. This microring resonator is similar to a Fabry–Perot resonator (see Fig. 4.2b), except that the microring is a traveling-wave resonator, while Fabry–Perot is a standing-wave resonator.

With regard to the transfer function equations, Eqs. 4.4–4.6, we substitute α and t by αt_2 and t_1, respectively, and then can obtain the transfer function at the throughput port, described by,

$$\frac{E_{t1}}{E_{i1}} = \frac{t_1 - \alpha t_2^* e^{j\theta}}{1 - \alpha t_1^* t_2^* e^{j\theta}} \tag{4.7}$$

At the drop port, the relation $E_{t2} = k_2 E_r e^{j\theta/2}$ exists.

According to Eq. 4.6, we can obtain

$$\frac{E_{t2}}{E_{i1}} = \frac{-k_1^* k_2 (\alpha e^{j\theta})^{1/2}}{1 - \alpha t_1^* t_2^* e^{j\theta}} \tag{4.8}$$

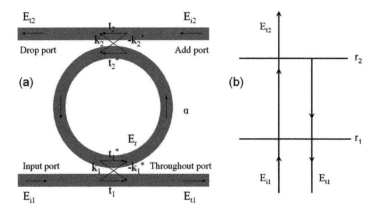

Figure 4.2 (a) Microring resonator coupled to two bus waveguides; (b) Fabry–Perot resonator.

Racetrack resonators are made with straight waveguides inserted into the coupling region to enhance the coupling efficiency between the racetrack and bus waveguides. When the coupling is very strong, the phase shift in the coupling region cannot be neglected. Thus,

$$t \to te^{j\varphi_t},\ t^* \to t^* e^{j\varphi_t},\ k \to ke^{-j\varphi_k},\ \text{and}\ k^* \to k^* e^{-j\varphi_k}.$$

Then, the transfer function at the throughput port and the drop port can be expressed as follows:

$$E_{t1} = \frac{t_1 e^{j\varphi_{t1}} - \alpha t_2^* e^{j\theta} e^{j\varphi_{t2}} (t_1 t_1^* e^{j2\varphi_{t1}} - k_1 k_1^* e^{-j2\varphi_{k1}})}{1 - \alpha t_1^* t_2^* e^{j\theta} e^{j\varphi_{t1}} e^{j\varphi_{t2}}} \quad (4.9)$$

$$E_{t2} = \frac{-\alpha^{1/2} k_1^* k_2 e^{j\theta/2} e^{-j\varphi_{k1}} e^{-j\varphi_{k2}}}{1 - \alpha t_1^* t_2^* e^{j\theta} e^{j\varphi_{t1}} e^{j\varphi_{t2}}} \quad (4.10)$$

4.2.2 Cascaded Microrings

When multiple microrings are cascaded [5, 6], the transmission becomes sophisticated. Here, the transfer matrix method has been employed to deduce the transfer functions of series-coupled microrings and parallel-coupled microrings.

4.2.2.1 Transfer matrix units

- **Directional coupler and transmission line**

For cascaded microrings, directional coupler and transmission line are basic elements having four ports, as shown in Fig. 4.3.

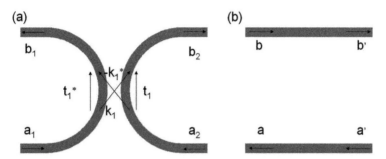

Figure 4.3 (a) directional coupler and (b) transmission line.

As seen in Fig. 4.3a, the transfer matrix equation for the directional coupler is expressed as

$$\begin{bmatrix} b_1 \\ b_2 \end{bmatrix} = \begin{bmatrix} t_1^* & -k_1^* \\ k_1 & t_1 \end{bmatrix} \begin{bmatrix} a_1 \\ a_2 \end{bmatrix}. \quad (4.11)$$

We can rewrite it as,

$$\begin{bmatrix} a_2 \\ b_2 \end{bmatrix} = \begin{bmatrix} 1/k_1 & -t_1/k_1 \\ t_1^*/k_1 & -1/k_1 \end{bmatrix} \begin{bmatrix} b_1 \\ a_1 \end{bmatrix}. \quad (4.12)$$

So, the transfer matrix P for directional coupler is expressed as

$$P = \begin{bmatrix} 1/k_1 & -t_1/k_1 \\ t_1^*/k_1 & -1/k_1 \end{bmatrix}. \quad (4.13)$$

Then, the transfer matrix equation for the transmission line in Fig. 4.3b is expressed as

$$\begin{bmatrix} b' \\ a' \end{bmatrix} = \begin{bmatrix} \alpha^{1/2} \exp(j\theta/2) & 0 \\ 0 & \alpha^{-1/2} \exp(-j\theta/2) \end{bmatrix} \begin{bmatrix} b \\ a \end{bmatrix}. \quad (4.14)$$

Also, the transfer matrix Q for the transmission line is expressed as

$$Q = \begin{bmatrix} \alpha^{1/2} \exp(j\theta/2) & 0 \\ 0 & \alpha^{-1/2} \exp(-j\theta/2) \end{bmatrix} \quad (4.15)$$

- **Microring resonator**

The microring resonator has four ports, as shown in Fig. 4.4. Using the transfer function for a microring coupled to two bus waveguides, when only a light injection in the input port or drop port is assumed, we can obtain the following expressions:

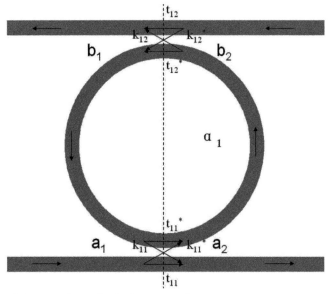

Figure 4.4 The schematic of a four-port microring resonator.

$$\tau_{11} = \frac{b_1}{a_1} = \frac{\alpha_1^{1/2} k_{11}^* k_{12} \exp(j\theta_1/2)}{1 - \alpha_1 t_{11}^* t_{12}^* \exp(j\theta_1)} \quad (4.16)$$

$$\tau_{12} = \frac{a_2}{b_2} = \frac{\alpha_1^{1/2} k_{11} k_{12}^* \exp(j\theta_1/2)}{1 - \alpha_1 t_{11}^* t_{12}^* \exp(j\theta_1)} \quad (4.17)$$

$$r_{11} = \frac{a_2}{a_1} = \frac{t_{11} - \alpha_1 t_{12}^* \exp(j\theta_1)}{1 - \alpha_1 t_{11}^* t_{12}^* \exp(j\theta_1)} \quad (4.18)$$

Principle of the Microring Resonator | 55

$$r_{12} = \frac{b_1}{b_2} = \frac{t_{12} - \alpha_1 t_{11}^* \exp(j\theta_1)}{1 - \alpha_1 t_{11}^* t_{12}^* \exp(j\theta_1)} \quad (4.19)$$

When the input port and the drop port both have a light injection, then,

$$b_1 = \tau_{11} a_1 + r_{12} b_2 \quad (4.20)$$

and

$$a_2 = \tau_{12} b_2 + r_{11} a_1. \quad (4.21)$$

Hence, we have

$$b_2 = \frac{1}{r_{12}} b_1 - \frac{\tau_{11}}{r_{12}} a_1 \quad (4.22)$$

and

$$a_2 = \frac{\tau_{12}}{r_{12}} b_1 + \frac{r_{11} r_{12} - \tau_{11} \tau_{12}}{r_{12}} a_1. \quad (4.23)$$

So, the transfer matrix equation for a microring resonator is expressed as

$$\begin{bmatrix} b_2 \\ a_2 \end{bmatrix} = \begin{bmatrix} 1/r_{12} & -\tau_{11}/r_{12} \\ \tau_{12}/r_{12} & (r_{11} r_{12} - \tau_{11} \tau_{12})/r_{12} \end{bmatrix} \begin{bmatrix} b_1 \\ a_1 \end{bmatrix}. \quad (4.24)$$

We denote the transfer matrix for the microring resonator as Y.

$$Y = \begin{bmatrix} 1/r_{12} & -\tau_{11}/r_{12} \\ \tau_{12}/r_{12} & (r_{11} r_{12} - \tau_{11} \tau_{12})/r_{12} \end{bmatrix} \quad (4.25)$$

4.2.2.2 Series-coupled microring resonators

Multiple microrings are series coupled in the configuration shown in Fig. 4.5.

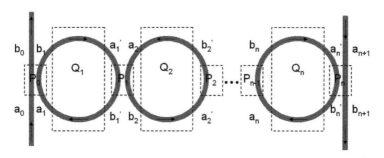

Figure 4.5 Series-coupled microring resonators.

56 | *Microring Resonators*

They can be divided into multiple directional couplers and transmission lines. Therefore, the transfer matrix equation for series-coupled microring resonators can be written as

$$\begin{bmatrix} b_{n+1} \\ a_{n+1} \end{bmatrix} = P_n Q_n P_{n-1} Q_{n-1} \cdots P_1 Q_1 P_0 \begin{bmatrix} b_0 \\ a_0 \end{bmatrix} = M \begin{bmatrix} b_0 \\ a_0 \end{bmatrix}, \tag{4.26}$$

where P_n and Q_n are, respectively, described as

$$\begin{cases} P_n = \begin{bmatrix} -t/k_n & 1/k_n \\ -1/k_n & -t^*/k_n \end{bmatrix} \\ Q_n = \begin{bmatrix} 0 & \alpha_n^{1/2} \exp(j\theta_n/2) \\ \alpha_n^{-1/2} \exp(-j\theta_n/2) & 0 \end{bmatrix} \end{cases} \tag{4.27}$$

The transfer matrix for cascaded microrings is written as

$$M = \begin{bmatrix} A & B \\ C & D \end{bmatrix}. \tag{4.28}$$

With no light injection at the add port, the transfer functions at the throughput port and the drop port can be expressed as follows:

$$\begin{cases} S_t = \dfrac{b_0}{a_0} = -\dfrac{D}{C} \\ S_d = \dfrac{b_{n+1}}{a_0} = B - \dfrac{AD}{C} \end{cases} \tag{4.29}$$

4.2.2.3 Parallel-coupled microring resonators

Figure 4.6 shows the configuration of parallel-coupled microring resonators, which can be divided into two elements, single-microring resonators and parallel transmission lines. Since there is no coupling between adjacent two microrings, the transfer matrix equation for a parallel-coupled microring can be written as

$$\begin{bmatrix} b_n \\ a_n \end{bmatrix} = Y_n X_{n-1} Y_{n-1} X_{n-2} \cdots Y_2 X_1 Y_1 \begin{bmatrix} b_0 \\ a_0 \end{bmatrix} = N \begin{bmatrix} b_0 \\ a_0 \end{bmatrix}, \tag{4.30}$$

where the transfer matrixes for the microring and the transmission line are expressed as

$$Y_n = \begin{bmatrix} 1/r_{n2} & -\tau_{n1}/r_{n2} \\ \tau_{n2}/r_{n2} & (r_{n1}r_{n2} - \tau_{n1}\tau_{n2})/r_{n2} \end{bmatrix} \tag{4.31}$$

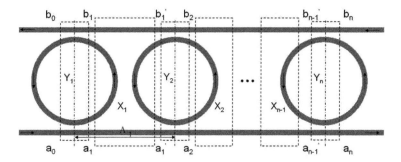

Figure 4.6 Parallel-coupled microring resonators.

and

$$X_n = \begin{bmatrix} \alpha_n^{1/2}\exp(j\theta_n') & 0 \\ 0 & \alpha_n^{-1/2}\exp(-j\theta_n') \end{bmatrix}, \quad (4.32)$$

where $\theta_n' = \beta\Lambda_n$.

The transfer matrix N for cascaded microrings is specified by

$$N = \begin{bmatrix} E & F \\ G & H \end{bmatrix}.$$

Therefore, the transfer function for the parallel-coupled microrings at the throughput port and the drop port can be expressed by:

$$S_t = \frac{a_n}{a_0} = H - \frac{FG}{E} \quad (4.33)$$

$$S_d = \frac{b_0}{a_0} = -\frac{F}{E} \quad (4.34)$$

Microring resonators can construct various functional devices with other optical components, for which transfer functions can be obtained using the transfer matrix method. First, we divide the complicated configuration into basic elements, such as directional coupler, transmission line, and microring resonator. Then, with their transfer matrixes employed, the transfer function for the aimed configuration can be derived.

4.3 Optical Properties of Microring Resonators

4.3.1 Properties of Amplitude

4.3.1.1 Single microring

- **Transmission spectrum**

Set the input optical intensity as 1; then the output optical intensities at the throughput port and the drop port are described as follows:

$$|E_{t1}|^2 = \frac{|t_1|^2 + \alpha^2 |t_2|^2 - 2\alpha |t_1 t_2| \cos(\theta + \varphi_{t1} + \varphi_{t2})}{1 + \alpha^2 |t_1 t_2|^2 - 2\alpha |t_1 t_2| \cos(\theta + \varphi_{t1} + \varphi_{t2})} \qquad (4.35)$$

and

$$|E_{t2}|^2 = \frac{\alpha |k_1 k_2|^2}{1 + \alpha^2 |t_1 t_2|^2 - 2\alpha |t_1 t_2| \cos(\theta + \varphi_{t1} + \varphi_{t2})}, \qquad (4.36)$$

where t_1 and t_2 are the transmitting coefficients and φ_{t1} and φ_{t2} are the phase shifts in the coupling-in region and coupling-out region, respectively. When $\theta + \varphi_{t1} + \varphi_{t2} = 2m\pi$ (m is an integer) is satisfied, the microring is resonating and also has

$$|E_{t1}|^2 = \frac{(|t_1| - \alpha |t_2|)^2}{(1 - \alpha |t_1 t_2|)^2} \qquad (4.37)$$

and

$$|E_{t2}|^2 = \frac{\alpha(1 - |t_1|^2)(1 - |t_2|^2)}{(1 - \alpha |t_1 t_2|)^2} \qquad (4.38)$$

Figure 4.7 plots the transmission spectrum of the microring resonator. The throughput port outputs most of the light in the case of off-resonance, while the drop port outputs most of the light in the case of on-resonance. At $|t_1| = \alpha |t_2|$, called the critical coupling condition, no light is transmitted to the throughput port and the optical intensity at the drop port achieves the maximum when the microring is resonating.

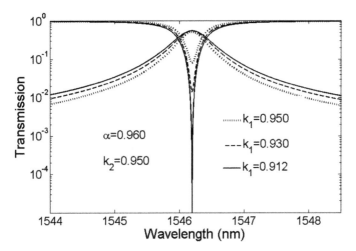

Figure 4.7 The transmission spectrum of the microring resonator.

- **Performance parameters**

The main parameters to characterize microring resonators are the FSR, quality factor, the extinction ratio (ER), and the intensity enhancement (IE) [7]. Figure 4.8 shows the transmission spectrum of the microring add-drop filter with these parameters labeled.

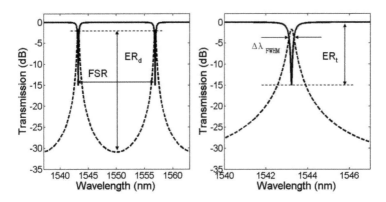

Figure 4.8 The transmission spectrum of a microring add-drop filter.

The FSR represents the spacing between two adjacent peaks in the transmission of a microring resonator, and it depends on the propagation constant dispersion.

$$\frac{\partial \beta}{\partial \lambda} = \frac{\partial (2\pi n_{\text{eff}} / \lambda)}{\partial \lambda} = \frac{\beta}{\lambda} + k \frac{\partial n_{\text{eff}}}{\partial \lambda} \tag{4.39}$$

Assume that the optical dispersion is weak; thus, we have $\partial \beta / \partial \lambda \approx -(\beta / \lambda)$ and then

$$\text{FSR} = -\frac{2\pi}{L} \left(\frac{\partial \beta}{\partial \lambda} \right)^{-1} \approx \frac{\lambda^2}{n_{\text{eff}} L} \tag{4.40}$$

However, this assumption brings a large error when the optical dispersion is strong. The group index n_g is specified by

$$n_g = \frac{\lambda}{k} \cdot \frac{\partial \beta}{\partial \lambda} = n_{\text{eff}} - \lambda \frac{\partial n_{\text{eff}}}{\partial \lambda} \tag{4.41}$$

So, the FSR is

$$\text{FSR} = -\frac{2\pi}{L} \left(\frac{\partial \beta}{\partial \lambda} \right)^{-1} \approx \frac{\lambda^2}{n_g L} \tag{4.42}$$

The quality factor Q is the ratio of power storage and power dissipation, and it is equal to the ratio of resonant wavelength and full width at half maximum. According to the transfer function of the microring add-drop filter, we can obtain

$$\left| \frac{-k_1^* k_2 (\alpha e^{j\theta})^{1/2}}{1 - \alpha t_1^* t_2 e^{j\theta}} \right|^2 = \frac{\alpha |k_1|^2 |k_2|^2}{2(1 - \alpha |t_1 t_2|)^2} . \tag{4.43}$$

There is no extra loss and a minor phase shift in the coupling region. Hence,

$$\frac{\alpha k_1^2 k_2^2}{1 - 2\alpha t_1 t_2 \cos(\Delta\theta) + (\alpha t_1 t_2)^2} = \frac{\alpha k_1^2 k_2^2}{2(1 - \alpha t_1 t_2)^2} . \tag{4.44}$$

When $\Delta\theta$ is very small, we can get $\cos(\Delta\theta) = 1 - \dfrac{(\Delta\theta)^2}{2}$ and $\Delta\theta = \dfrac{(1 - \alpha t_1 t_2)^2}{\alpha t_1 t_2}$.

The quality factor can be expressed as

$$Q = \frac{\omega_0}{2\Delta\omega} = \frac{(2\pi c / \lambda_0)}{2\Delta\theta c / (n_{\text{eff}} L)} = \frac{\pi n_{\text{eff}} L}{\lambda_0 \Delta\theta} . \tag{4.45}$$

Therefore, the quality factor is

$$Q = \frac{\pi n_{\text{eff}} L}{\lambda_0} \cdot \frac{\sqrt{\alpha t_1 t_2}}{1 - \alpha t_1 t_2}. \quad (4.46)$$

The ER determines the depth of resonant peaks, which is equal to the ratio of maximum power and minimum power at the output port.

The output power at the throughput port is

$$\left| E_{t1} \right|^2 = \frac{t_1^2 + \alpha^2 t_2^2 - 2\alpha t_2 \cos\theta}{1 + \alpha^2 t_1^2 t_2^2 - 2\alpha t_1 t_2 \cos\theta}. \quad (4.47)$$

Then, the extinction ratio is

$$ER_t = \frac{\left| E_{t1} \right|^2_{\max}}{\left| E_{t1} \right|^2_{\min}} = \frac{\left| E_{t1} \right|^2_{\cos\theta=1}}{\left| E_{t1} \right|^2_{\cos\theta=-1}} = \frac{(t_1 - \alpha t_2)^2}{(1 - \alpha t_1 t_2)^2} - \frac{(1 + \alpha t_1 t_2)^2}{(t_1 + \alpha t_2)^2}. \quad (4.48)$$

Similarly, we can get the extinction ratio at the drop port

$$ER_d = \frac{(1 + \alpha t_1 t_2)^2}{(1 - \alpha t_1 t_2)^2} \quad (4.49)$$

The IE is the ratio of optical power in the resonator and that in the bus waveguides. For the microring add-drop filter, the optical power in the microring satisfies

$$\frac{E_{t2}}{E_{i1}} = \frac{-k_1^*}{1 - \alpha t_1^* t_2^* e^{j\theta}}. \quad (4.50)$$

The power in the microring is considered to be uniform; then, the intensity enhancement is

$$\text{IE} = \left| \frac{E_{t2}}{E_{i1}} \right|^2_{\phi=0} = \frac{k^2}{1 + \alpha^2 t_1^2 t_2^2 - 2\alpha t_1 t_2}. \quad (4.51)$$

When $k_1 = k_2 = k$ and $k < 1$, we have IE $\approx k^2$.

4.3.1.2 Cascaded microrings

- **Optical filter with a box-like response**

The ideal response is shaped like a box for optical filters used in the wavelength division multiplexing system. The Butterworth

62 | *Microring Resonators*

filter has the flattest pass-band and stop-band, while the Chebyshev filter has the sharpest band edge and small waviness in the pass-band or the stop-band [2]. With some conditions satisfied, cascaded microrings have a better response than a single microring. Table 4.1 gives the necessary conditions for cascaded microrings to construct a Butterworth filter and a Chebyshev filter.

Table 4.1 The necessary conditions for Butterworth and Chebyshev filters

N	Butterworth filter	Chebyshev filter
2	$\mu_1^{*2} = 0.250\mu^{*4}$	$\mu_1^{*2} = 0.250\mu^{*4}(1+2y)$
3	$\mu_1^{*2} = \mu_1^{*2} = 0.125\mu^{*4}$	$\mu_1^{*2} = \mu_2^{*2} = 0.125\mu^{*4}(1+1.5y^{2/3})$
4	$\mu_1^{*2} = \mu_3^{*2} = 0.100\mu^{*4}$ $\mu_2^{*2} = 0.040\mu^{*4}$	
5	$\mu_1^{*2} = \mu_4^{*2} = 0.0955\mu^{*4}$ $\mu_2^{*2} = \mu_3^{*2} = 0.040\mu^{*4}$	
6	$\mu_1^{*2} = \mu_5^{*2} = 0.0915\mu^{*4}$ $\mu_2^{*2} = \mu_4^{*2} = 0.0245\mu^{*4}$ $\mu_3^{*2} = 0.0179\mu^{*4}$	

The waviness in the pass-band of the Chebyshev filter is determined by y in Table 4.1, where μ_n and μ are

$$\mu_n^2 = k_n^2 \frac{v_{gn}v_{gn+1}}{(2\pi R_n)(2\pi R_{n+1})} \tag{4.52}$$

and

$$\mu^2 = k^2 \frac{v_{g1}}{2\pi R_1}. \tag{4.53}$$

In the expressions above, k_n is the coupling coefficient between ring n and ring $n + 1$, k is the coupling coefficient between ring and bus waveguide, and R_n and v_{gn} are the radius and group velocity of the guided mode in ring n. Figure 4.9 plots the transmission of a single ring, cascaded double rings, and cascaded triple rings. One notices that the pass-band becomes flat and the band edge becomes sharp, as the number of microrings increases.

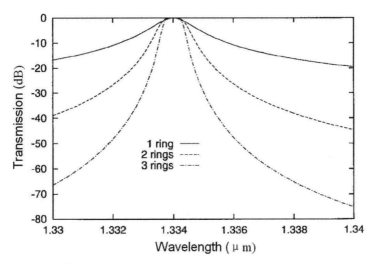

Figure 4.9 The transmissions of a single ring and cascaded microrings.

- **The Vernier effect**

The cascaded microring resonators show the Vernier effect [8] when their radii are different.

Take coupled double microrings as an example, shown in Fig. 4.10a; the light tunnels to the output port only when both microrings are resonating. Hence, the FSR of double-coupled microrings satisfies

$$\text{FSR}_{12} = N_1 \cdot \text{FSR}_1 = N_2 \cdot \text{FSR}_2, \quad (4.54)$$

where FSR_1, FSR_2, and FSR_{12} are the FSRs of ring 1, ring 2, and coupled double rings, respectively. Both N_1 and N_2 are integers. Figure 4.10b shows the output transmissions of the individual ring and coupled rings. We find that the FSR is enlarged when the ring radii are different.

64 | Microring Resonators

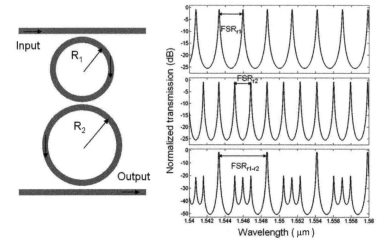

Figure 4.10 (Left) Series-coupled double microrings with different radii and (right) the Vernier effect of cascaded microrings. © 2002 IEEE. Reprinted, with permission, from Ref. [8].

4.3.2 Properties of Phase

As the optical phase is sensitive to wavelength, the light would be slowed or accelerated by microring resonators [9]. Therefore, it is promising for a microring resonator to act as an optical buffer in an optical interconnect system.

The optical phase from an all-pass filter based on a single microring is expressed as

$$\frac{E_t}{E_i} = \exp[j(\pi+\theta)]\frac{\alpha - t\exp(-j\theta)}{1-\alpha t\exp(j\theta)} \quad (4.55)$$

and

$$\Phi = \arg\left(\frac{E_t}{E_i}\right)$$
$$= \pi + \theta + \arctan\left(\frac{t\sin\theta}{\alpha - t\cos\theta}\right) + \arctan\left(\frac{\alpha t\sin\theta}{1-\alpha t\sin\theta}\right). \quad (4.56)$$

For an add-drop filter based on a single microring, the throughput optical phase is

$$\frac{E_{t1}}{E_{i1}} = \exp[j(\pi+\theta)]\frac{\alpha t_2 - t_1\exp(-j\theta)}{1-\alpha t_1 t_2\exp(j\theta)} \quad (4.57)$$

and

$$\Phi = \arg\left(\frac{E_{t1}}{E_{i1}}\right)$$
$$= \pi + \theta + \arctan\left(\frac{t_1 \sin\theta}{\alpha t_2 - t_1 \cos\theta}\right) + \arctan\left(\frac{\alpha t_1 t_2 \sin\theta}{1 - \alpha t_1 t_2 \cos\theta}\right) \quad (4.58)$$

and the drop optical phase is

$$\frac{E_{t2}}{E_{i1}} = \frac{k_1 k_2 [\alpha \exp(i\theta)]^{1/2}}{1 - \alpha t_1 t_2 \exp(i\theta)} \quad (4.59)$$

and

$$\Phi = \arg\left(\frac{E_{t2}}{E_{i1}}\right) = \frac{\theta}{2} + \arctan\left(\frac{\alpha t_1 t_2 \sin\theta}{1 - \alpha t_1 t_2 \cos\theta}\right). \quad (4.60)$$

The curves in Fig. 4.11 show the optical phases from the microring add-drop filter as a function of the wavelength when there is undercoupling ($t_1 > t_2$), symmetric coupling ($t_1 = t_2$), or overcoupling ($t_1 < t_2$). The resonance wavelength is chosen at about 1.55 μm. At the throughput port, the phase jumps at the resonance frequency whereas the phase increases monotonically at the drop port.

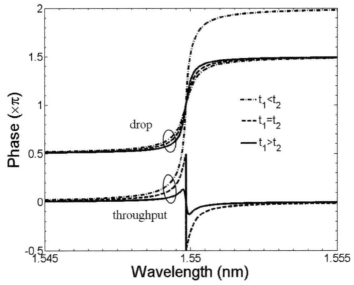

Figure 4.11 The phase curves at the throughput port and the drop port for a microring add-drop filter.

To get slow light or fast light, we have to change the optical group velocity through structure dispersion or material dispersion. The group velocity of light is described by

$$v_g = \frac{\partial \omega}{\partial k} = \frac{c - \omega \frac{\partial n(\omega,k)}{\partial k}}{n(\omega,k) + \omega \frac{\partial n(\omega,k)}{\partial \omega}}. \tag{4.61}$$

A microring resonator can slow down the light through structure dispersion. As seen in Fig. 4.12, the optical pulse of the Gaussian shape is delayed for picoseconds, reflecting that the microring resonator captures, stores, and then releases the light. Note that the pulse width should be comparable with the lifetime in the resonator, as otherwise, the output pulse would be distorted.

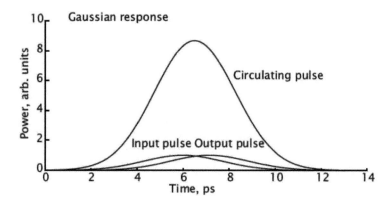

Figure 4.12 The time delay of the optical pulse in the Gaussian shape.

The time delay is denoted by

$$\tau_d = -\frac{d\Phi}{d\theta} \cdot \frac{d\theta}{d\omega} \quad \text{and} \quad \frac{d\theta}{d\omega} = \frac{n_g L}{c}. \tag{4.62}$$

For a microring all-pass filter

$$\frac{d\Phi}{d\theta} = \frac{\alpha^2 - \alpha t \cos\theta}{\alpha^2 + t^2 - 2\alpha t \cos\theta} + \frac{\alpha t \cos\theta - \alpha^2 t^2}{1 + \alpha^2 t^2 - 2\alpha t \cos\theta}, \tag{4.63}$$

Therefore,

$$\tau_d = -\left(\frac{\alpha^2 - \alpha t \cos\theta}{\alpha^2 + t^2 - 2\alpha t \cos\theta} + \frac{\alpha t \cos\theta - \alpha^2 t^2}{1 + \alpha^2 t^2 - 2\alpha t \cos\theta}\right) \cdot \frac{n_g L}{c} \tag{4.64}$$

For the throughput port of the microring add-drop filter

$$\frac{d\Phi}{d\theta} = \frac{\alpha^2 t_2^2 - \alpha t_1 t_2 \cos\theta}{\alpha^2 t_2^2 + t_1^2 - 2\alpha t_1 t_2 \cos\theta} + \frac{\alpha t_1 t_2 \cos\theta - \alpha^2 t_1^2 t_2^2}{1 + \alpha^2 t_1^2 t_2^2 - 2\alpha t_1 t_2 \cos\theta} \quad (4.65)$$

and

$$\tau_d = -\left(\frac{\alpha^2 t_2^2 - \alpha t_1 t_2 \cos\theta}{\alpha^2 t_2^2 + t_1^2 - 2\alpha t_1 t_2 \cos\theta} + \frac{\alpha t_1 t_2 \cos\theta - \alpha^2 t_1^2 t_2^2}{1 + \alpha^2 t_1^2 t_2^2 - 2\alpha t_1 t_2 \cos\theta}\right) \cdot \frac{n_g L}{c}. \quad (4.66)$$

For the drop port of the microring add-drop filter

$$\frac{d\Phi}{d\theta} = \frac{1}{2} + \frac{\alpha t_1 t_2 \cos\theta - \alpha^2 t_1^2 t_2^2}{1 + \alpha^2 t_1^2 t_2^2 - s2\alpha t_1 t_2 \cos\theta} \quad (4.67)$$

and

$$\tau_d = -\left(\frac{1}{2} + \frac{\alpha t_1 t_2 \cos\theta - \alpha^2 t_1^2 t_2^2}{1 + \alpha^2 t_1^2 t_2^2 - 2\alpha t_1 t_2 \cos\theta}\right) \cdot \frac{n_g L}{c}. \quad (4.68)$$

We can conclude from Fig. 4.13 that slow light or fast light achieves the maximum at the resonant wavelength. When it is undercoupling ($t_1 > t_2$), light is speeded at the throughput port, while it is slowed at the drop port. When it is symmetrical coupling ($t_1 = t_2$), no light is observed at the throughput port, while light is slowed at the drop port. When it is overcoupling ($t_1 < t_2$), light is slowed both at the throughput port and the drop port.

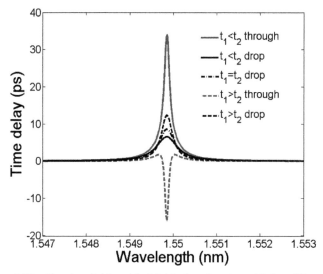

Figure 4.13 The slow light and fast light of a microring add-drop filter.

4.4 Design of Microring Resonators

Microring resonators can be characterized by transfer functions, which are mainly influenced by guided modes in waveguide, coupling efficiency, and internal loss. First, we calculate such parameters for the designed structure and then substitute them into the transfer function; thus, we can simulate the response of the microring resonator, which is much faster than a full numerical method. In this section, we will introduce the design course for including submicron waveguide, directional coupler, and internal loss in the microring.

4.4.1 Waveguide Design

- **Single-mode condition**

The waveguides in microring resonators should have a single guided mode. The properties of guided modes are determined by the waveguide structure, such as width, height, and etching depth. As dimensions are scaled down, the single-mode condition for submicron waveguides becomes different from the traditional single-mode condition [10], as shown in Fig. 4.14. It is found that rib waveguides still support a single mode with the ratio of rib height

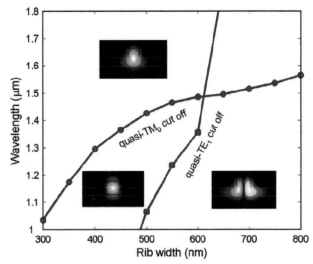

Figure 4.14 The mode cut-off condition for submicron SOI waveguides. Reprinted from Ref. [10], Copyright (2009), with permission from Elsevier.

to slab height lower than 0.5. Moreover, the transverse-magnetic mode is no longer guided in the waveguide of shallow-etched ribs since it is more likely to leak from the slab as the waveguide width decreases. The properties of bent waveguides are similar to those of straight waveguides. However, the mode profile lies outside the middle of the waveguide and radiation loss cannot be neglected for bent waveguides.

- **Polarization independence**

The index and stress profiles are anisotropic for waveguides with an asymmetric cross section, making rather large differences between transverse-electric mode and transverse-magnetic mode, especially for submicron waveguides, as seen in Fig. 4.15. The main solutions for reducing polarization dependence are:
  - o Optimizing the etching depth and waveguide width to realize polarization independence
  - o Tuning the thickness and strain of oxide cladding to realize polarization independence by introducing material stress

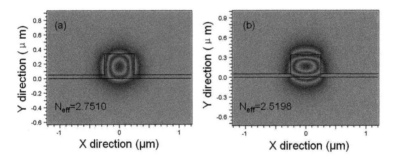

Figure 4.15 The TE mode (a) and TM mode (b) of a rib waveguide.

However, both these methods demand high-accuracy fabrication techniques. Another method is using polarization converters to make the subsystem polarization entirely independent while the microring resonators are still polarization dependent partially [11].

4.4.2 Coupler Design

The microring and microring/straight waveguide are usually coupled using a directional coupler or multimode interferometer

(MMI). The fabrication tolerance for the MMI is relatively large, while the splitting ratio is inflexible and the insertion loss is large. On the contrary, the splitting ratio is flexible and the insertion loss is low for a directional coupler. Here, we focus on the directional coupler.

The spacing is changing for the bent coupling [12], as shown in Fig. 4.16, where R denotes the ring radius and the gap denotes the minimum spacing. Suppose the dimensions of coupled waveguides are identical, and they are far smaller than the ring radius. Then, the phase mismatch and the optical loss in the coupling region can be neglected. Therefore, the transmitting coefficient t and coupling k are expressed as follows:

$$t = \cos\left(\int_{-L}^{L} k(z)dz\right), \quad k = \sin\left(\int_{-L}^{L} k(z)dz\right), \tag{4.69}$$

where $k(z)$ is the coupling coefficient as a function of position in the z direction. As the coupling is angled, it should be revised by $k(z) = k_{||}(z) \cos \theta$, where $k_{||}(z)$ is the parallel coupling coefficient and it is given by

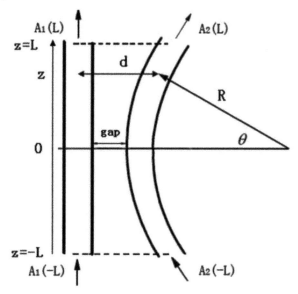

Figure 4.16 The coupling between a bent waveguide and a straight waveguide.

$$k_{\|}[d(z)] = \frac{i\omega\varepsilon_0}{4} \int_{-\infty}^{\infty} (n_1^2 - n_2^2) E_1^*(x) \cdot E_2(x) dx, \quad (4.70)$$

where n_1 and n_2 are indexes of core and cladding, respectively; $E_1^*(x)$ and $E_2(x)$ are the optical field profiles in two waveguides; ω is the optical angle frequency; and ε_0 is the dielectric constant in vacuum.

To deal with the coupling of 3D waveguides, such as rib waveguides and strip waveguides, we have to transform them into 2D slab waveguides using the effective index method at first. The coupling efficiency is influenced by the following factors: rib width, rib height, slab height, bending radius, and spacing gap. As shown in Fig. 4.17, the coupling efficiency can be enhanced by decreasing the rib width/height ratio or increasing the slab height, which may weaken the optical constraint in waveguides. By narrowing the spacing, enlarging the bending radius, or lengthening the coupling region, we can enhance the coupling greatly. For a given geometry, it is found that the coupling efficiency increases with wavelength.

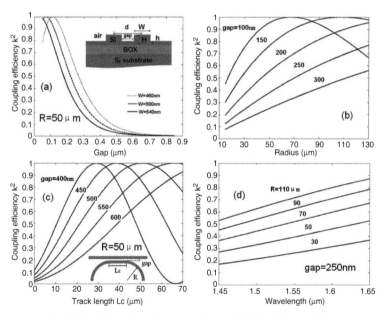

Figure 4.17 Coupling of ridge waveguides. (a) Coupling efficiency versus gap; inset: the cross section of the coupling region; (b) coupling efficiency versus bending radius; (c) coupling efficiency versus track length; (d) coupling efficiency versus wavelength.

4.4.3 Internal Loss

The absorption coefficient is very small for bulk silicon, and generally, the optical nonlinear effects are very weak. Therefore, the total optical absorption can be neglected in SOI waveguides. In reality, the internal loss mainly comes from bending loss and scattering loss.

- **Bending loss**

A simple method to calculate bending loss is proposed by Marcuse [13], from which we first transform the 3D waveguides into slab waveguides using the effective index method and then calculate the bending loss based on slab waveguides by using the following expressions [14]:

$$\alpha_{bend} = \frac{\alpha_y^2}{k_0^3(1+\alpha_y w/2)(n_{e2}^2 - n_{e1}^2)} \frac{k_y^2}{} \exp(\alpha_y w) \exp\left(\frac{-2\alpha_y^3}{3n_e^2 k_0^2} R\right) \quad (4.71)$$

and

$$Loss_{bend} = -10\log_{10}(\exp(-\alpha_{bend}\Delta\theta R)) \quad (4.72)$$

where $\alpha_y = k_0(n_e^2 - n_{e1}^2)^{1/2}$, $k_y = k_0(n_{e2}^2 - n_e^2)^{1/2}$, $k_0 = 2\pi/\lambda_0$, and α_{bend} are the imaginary parts of the propagation constant, denoting the bending loss, n_e is the effective index of the fundamental mode, n_{e1} is the effective index of the outer slab region, and n_{e2} is the effective index of the inner slab region.

Figure 4.18 shows the bending loss of the 90° curvature for SOI rib waveguides, with a rib height of 0.34 µm and an etching depth of

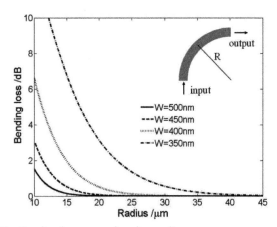

Figure 4.18 Bending loss versus bending radius.

0.17 μm, according to Eqs. 4.71–4.72. It is obvious that the bending loss rises rapidly with the bending radius and almost no bending loss exists for bending radii larger than 40 μm.

Bending loss is influenced not only by radius but also by the optical constraint in waveguides. Generally, the smaller the mode profile, the smaller is the radiation loss. Therefore, an effective method to reduce bending loss is enlarging the bending radius or increasing the etching depth.

- **Scattering loss**

The internal total reflecting model [15, 16] can explain the scattering loss. The roughness at the boundaries causes optical scattering when the light reflects between the upper and lower interfaces as propagating. The rougher the surface, the larger is the scattering loss, especially when the intensity at the interface is large.

On the basis of the Raleigh scattering theory, Tien presented an optical scattering model [16], suitable for asymmetric slab waveguides, and the power attenuation can be evaluated by

$$\alpha = K^2 \left(\frac{\cos^3 \theta}{2\sin\theta} \right) \left\{ \frac{1}{[t + (1/p_{10}) + (1/p_{12})]} \right\}, \tag{4.73}$$

where $K = (4\pi/\lambda)(\sigma_{12}^2 + \sigma_{12}^2 + \sigma_{10}^2)^{1/2}$, θ is the light transmitting angle, λ is the optical wavelength in waveguides, t is the waveguide thickness, and $1/p_{10}$ and $1/p_{12}$ are the depths of light in cladding and substrate, respectively. In addition, σ_{10} and σ_{12} are the mean square roots of roughness in cladding and substrate, respectively.

The scattering loss is proportional to the roughness and the square of index contrast at the boundaries. It is found that the scattering loss mainly comes from the roughness at sidewalls. Therefore, it is very essential to reduce roughness by optimizing lithography and dry etching techniques for submicron SOI waveguides.

4.5 Fabrication and Measurement of Microring Resonators

SOI material is compatible with advanced CMOS processes. Specific SOI processes have been applied in microelectronics for a long time in order to improve bulk insulation and decrease parasitic capacitances. SOI is frequently employed for optical waveguides, and the light is restricted and propagating in the top silicon layer. In this section, we will introduce the key fabrication techniques as well as the optical measurement of photonic devices.

4.5.1 Fabrication Flow

Figure 4.19 shows a typical scheme of the fabrication flow for a microring resonator. In the given scheme electron beam lithography (EBL) and dry etching (inductively coupled plasma [ICP]) are used for pattern definition. First, clean the SOI wafer. Then, spin the electron beam resist onto the wafer and bake it on a hot plate to make it hard. After that, transfer the mask pattern to the resist by EBL and etch the pattern onto the Si top layer. Remove the remaining resist, and the basic fabrication of the device is finished. Oxide cladding is often grown on the top to protect the waveguides.

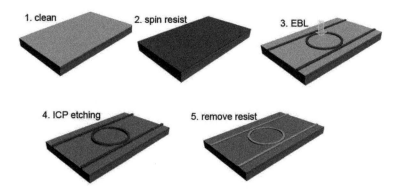

Figure 4.19 The fabrication flow of a microring resonator.

Table 4.2 lists some published processing variants and experimental results. Deep ultraviolet (DUV) lithography (commonly the krypton fluoride laser at 248 nm wavelength and the argon fluoride laser at 193 nm wavelength) and i-line lithography (365 nm wavelength from a Hg lamp) are both standard microelectronic process steps with large batch production, high speed, and low cost. On a laboratory scale, EBL and nanoimprint are flexible tools to realize optical devices of submicron dimensions.

Table 4.2 Fabrication methods for microring resonators

	Lithography	Etching	Cladding	Loss (dB/cm)	Q factor	FSR: nm
HKUST [17]	i-line	RIE	oxide	20	10,000	—
Ghent [18]	DUV	RIE	oxide	2.4	3000	14
IBM [19]	EBL	RIE	air	1.7	14,000	11
Cornell [20]	EBL	ICP	oxide	—	39,350	15
Purdue [21]	EBL	ICP	air	2–3	62,000	16
Aachen [22]	Nano-Imprint	RIE	air	3.5	47,700	8.3

4.5.2 Electron Beam Lithography

The wavelength of a high-energy electron wave is very short. Thus, EBL has the potential of very high resolution. EBL is widely used for scientific research due to the high accuracy. The fabrication results are mainly influenced by EB resist, proximity effect, write-field stitching error, and developing and fixing conditions.

- **Electron beam resist**

Electron beam resist belongs to polymers, chains of which can be reassembled under electron beam exposure. The exposed/unexposed region can be dissolved by a developing and fixing solution, with the mask pattern left.

- **Proximity effect**

Because of electron backscattering, electrons may spread to the near edges of adjacent patterns. And in the pattern, backscattering makes the electron density at the boundaries less than that at the center, which results in reshaping of the designed structure, as seen in Fig. 4.20a. To surpass the influence of proximity effects, pattern compensation or dose compensation methods are necessary.

- **Write-field stitching**

A large-area pattern is composed of many small write-fields for EBL. The scanning direction may be divergent from the moving direction of the sample platform, and that may result in a stitching error, as shown in Fig. 4.20b. The write-field stitching error cannot be eliminated easily, so the design would better lay important parts of patterns in a single write-field and far from the write-field borders.

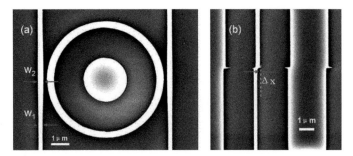

Figure 4.20 EBL: (a) proximity effect in the coupling region and (b) write-field stitching error.

- **Developing and fixing**

The dissolvability of resists changes a lot in the developer after the resists are exposed to an electron beam. Fixing is used to remove the remaining developer. The developer, fixer, and developing and fixing conditions (such as time and temperature) should be correctly chosen. Otherwise, the resist pattern misses the design targets.

4.5.3 Dry Etching Process

As an etching technique, wet etching is mostly isotropic and causes undercut. Using special solutions, wet etching can be anisotropic, but it is quite dependent on orientation. Therefore, dry etching is widely employed to fabricate submicron waveguides, which is anisotropic etching and able to get steep sidewalls.

ICP etching is a popular dry etching technique using a chemical reaction and plasma striking to remove the unprotected material, as shown in Fig. 4.21. Here, SF_6 and C_4F_8 are chosen as gases to etch silicon and passivate etching sidewalls, respectively. Both are very reactive gases that are partly ionized to a plasma by electrons accelerated under an applied voltage.

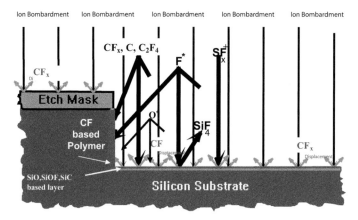

Figure 4.21 The dry etching of silicon by a plasma discharge.

The vacuum pump will extract the gas reaction products. Due to the high density of the plasma and low pressure in the ICP etching system, a high depth-to-width ratio and a high etching speed are obtained. The methods to enhance anisotropy are listed as follows:

- Increase the radio frequency (RF) power, which enhances the plasma striking, resulting in weak lateral etching.
- Lower the pressure in the reaction room, which increased the plasma free path, causing less scattering collision of the ions.
- Decrease the temperature of the sample surface, which reduces the chemical etching, as well as the lateral etching.

The RF power, pressure, constituent, and flow rate of etching gases in the reaction room play an important role in the etching process. Figure 4.22 shows the etching results of SOI-based microring resonators.

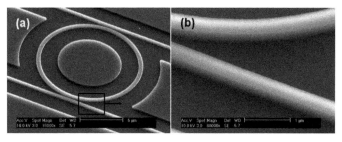

Figure 4.22 The SEM images of a microring resonator fabricated by ICP etching.

4.5.4 Silicon Oxide Growing

Thermal oxidation and chemical vapor deposition are the two main methods for silicon oxide growing. The speed of thermal oxidation is low, but this process causes less roughness and reduces optical propagation loss. SiO_2 thickness increases as the silicon is thermally oxidized. As seen in Fig. 4.23, the oxidation speed is high for juts at the boundaries of Si and SiO_2. Thus, the etching surfaces of Si are flattened after the oxide is removed. Xia et al. reduced the propagation loss in submicron SOI waveguides to 1.7 dB/cm using two-step thermal oxidation [18].

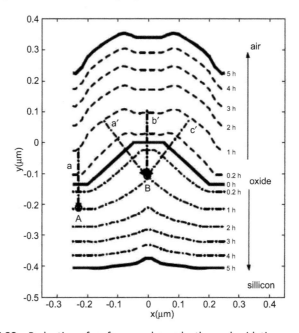

Figure 4.23 Reduction of surface roughness by thermal oxidation.

Silicon oxide can also be obtained by plasma-enhanced chemical vapor deposition (PECVD) or low-pressure chemical vapor deposition (LPCVD) with growth gases of SiH_4 and O_2. PECVD has a high deposition speed and thick oxide, while LPCVD has a dense oxide film and strong gap-filling ability. Figure 4.24 shows silicon oxide deposited by PECVD.

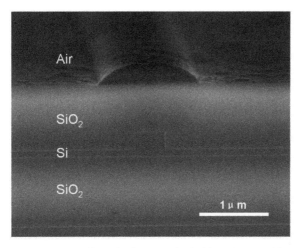

Figure 4.24 SOI waveguide with silicon oxide cladding deposited by PECVD.

4.5.5 Measurement

To measure the optical properties of microring resonators, such as filtering and time delay, we employ end fire coupling between fiber and device on the input port. A lens and a polarizer are placed behind the output port to collimate the light beam with a determined polarization, which is then focused and collected by a single-mode fiber with a collimator at the end.

- **Filtering measurement**

Figure 4.25 plots the setup for filtering measurement. The light from a frequency-tunable laser is coupled into the waveguides through a polarization controller and a polarization-maintaining fiber. The output light is collimated by a lens and then filtered by a polarizer. By turning the mirror, we can switch the collimated light either into an infrared camera or into a fiber with a collimator at the front end. With the light spot observed on the camera, we tune the measurement stages for accurate coupling. After that, the mirror is turned to lead the light to an optical spectrum analyzer (OSA) synchronously and then the transmission spectrum on the OSA is displayed.

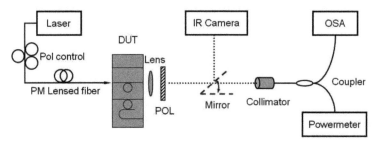

Figure 4.25 The measurement setup for microring filter characterization.

- **Time delay measurement**

As shown in Fig. 4.26, this is a setup for time delay measurement. The modulated optical signal is split into two parts, one going through the resonators and the other for reference. The main signal will be chosen to detect the optical switch at the output port. Detect and store the reference light and light from devices in oscillograph, respectively. Then, we can obtain the time delay by comparing these two waveforms.

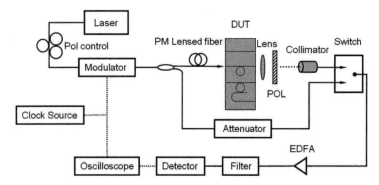

Figure 4.26 The measurement setup for time delay of microring resonators.

4.6 Applications of Microring Resonators

Microring resonators serve as building blocks in various functional components of integrated photonics. First, because of wavelength selectivity, a microring resonator can be used as an optical filter that is even tunable by thermo-optic effect or electro-optic effect. Second, the optical intensity in the microring is enhanced greatly compared to the input waveguide and, hence, the optical nonlinearity becomes significant in the microring resonator, leading to novel devices, such as Raman lasers and optical logic devices. Third, the optical pulse is time delayed due to wave dispersion in the microring resonator, enabling optical buffers in optical interconnects. In this section, we will illustrate the applications of SOI-based microring resonators, mainly in these three aspects.

4.6.1 Optical Filter

Microring resonators are wavelength sensitive and traveling-wave resonating, which is promising for optical filtering. They can construct many kinds of filters, such as all-pass filter, add-drop filter, optical demultiplexer/multiplexer, and tunable filter.

- **All-pass filter**

The microring resonator coupled to a bus waveguide is an all-pass filter, for which the transfer function is calculated from Eq. 4.4 using Eqs. 4.69 and 4.70 for determining the transfer coefficients.

Lowering the coupling efficiency and increasing the round-trip attenuation would decrease the bandwidth of the stop-band. The extinction ratio is determined by the relation of coupling efficiency and round-trip attenuation, which achieves maximum at the critical coupling. Figure 4.27 shows the microscope image and transmission spectrum of an all-pass filter based on an SOI microring resonator [23].

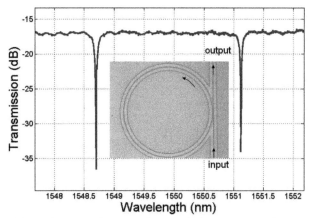

Figure 4.27 Microscope image and transmission spectrum of a an SOI-based optical filter.

Seen in Fig. 4.28, with a microring coupled to an arm, the Mach–Zehnder interferometer (MZI) can also act as an all-pass filter [24], for which the transfer function is expressed by the interference of both arms, one with and one without a coupled microring.

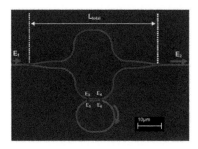

Figure 4.28 All-pass filter based on MZI coupled with a microring resonator. Reprinted with permission from Ref. [24] © The Optical Society.

The lengths of two arms are the same in the given example of Fig. 4.28. Therefore, the phase difference is $2m\pi$ (m is an integer) for off-resonance, while it is $(2m + 1)\pi$ for on-resonance, which causes destructive interference. The phase difference between both arms has to be considered if the arm lengths are not equal. By changing coupling efficiency, the bandwidth of the all-pass resonator may be tuned effectively.

- **Add-drop filter**

In the wavelength division multiplexing system, the add-drop filter is a key component, which adds or drops signals of special wavelengths at the network nodes. A microring resonator coupled with two bus waveguides can serve as an add-drop filter [25].

Figure 4.29 shows the SEM image and transmission spectrum of an add-drop filter based on a microring resonator. It is band-stop filtering and band-pass filtering at the throughput port and the drop port, respectively. Moreover, we find that the microring add-drop filter is polarization dependent and the transmission is quite different for TE and TM polarization. Therefore, the microring add-drop filter can be regarded as a polarization splitter, and here the splitting ratio of TE and TM polarization exceeds 20 dB.

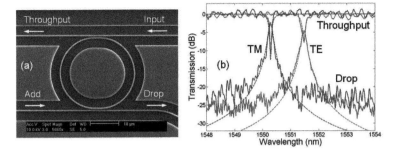

Figure 4.29 The SEM image and transmission spectrum of a microring add-drop filter. Reproduced from Ref. [25] (https://iopscience.iop.org/article/10.1088/1464-4258/11/1/015506/meta). © IOP Publishing. Reproduced with permission. All rights reserved.

The ideal response for filters is box-like, which means flat pass-band, and steep band edge. Cascades of multiple microrings in a series configuration [26], with a small footprint of only 7×10^{-4} mm^2, improve the filter response. The pass-band of five microrings is more box-like than that of three microrings and the out-of-band rejection is larger (40 dB).

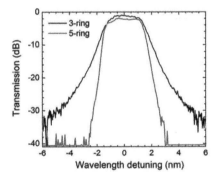

Figure 4.30 Transmission spectrum of an add-drop filter based on multiple microrings. Reprinted with permission from Ref. [26] © The Optical Society.

- **Demultiplexer/multiplexer**

A demultiplexer/multiplexer can be constructed by combining multiple add-drop filters, each corresponding to a single channel. There are several guidelines for the design. The FSR of every add-drop filter should be larger than the total bandwidth of all channels.

The spacing between adjacent two channels should be uniform. To enable low cross talk, the out-of-band rejection should be large enough for each add-drop filter.

Figure 4.31 plots an eight-channel demultiplexer in silicon [27]. The add-drop filters are based on series-coupled double microring resonators, with the drop port and the throughput port in the same direction, high out-of-band rejection, and low cross talk between channels. From the transmission, the out-of-band rejection is as large as 40 dB and the cross talk is lower than −30 dB.

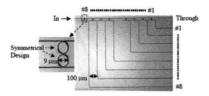

Figure 4.31 Eight-channel wavelength demultiplexer based on microring resonators. Reprinted with permission from Ref. [27] © The Optical Society.

- **Tunable filter**

A tunable filter is realizable by using a thermo-optic effect or a free carrier dispersion effect in silicon. The relation between index change and wavelength shift is described by

$$\Delta\lambda = \lambda_0 \Delta n_{\text{eff}}/n_{\text{eff}} \tag{4.74}$$

For silicon at room temperature, the thermo-optic coefficient is

$$dn/dT = 1.84 \times 10^{-4}/\text{K} \tag{4.75}$$

Thermo-optic tuning has many merits, such as fabrication simplicity, high modulation, and low power consumption, whereas the response time is limited to microsecond levels. For a high-speed response, we need to employ the free carrier dispersion effect in silicon.

At the wavelength of 1.55 μm, the index change caused by free carrier dispersion effect is

$$\Delta n = -[8.8 \times 10^{-22} \times \Delta N_e + 8.5 \times 10^{-18} \times (\Delta N_h)^{0.8}], \tag{4.76}$$

where ΔN_e and ΔN_h denote the concentration changes of electron and hole in units of cm^{-3}, respectively.

The light intensity or routing will be changed when a voltage is applied. Hence, tunable filters are applicable for optical modulators/ switches due to the same working mechanism. The output power spectrum is sensitive to a voltage sweep. The optical power is switched on/off within a narrow bandwidth. The speed of the modulator/switch can be increased by shortening the carrier transmitting distance and using a pre-emphasis rectangular pulse, as proposed and demonstrated by Xu et al., with a data transmission rate of 12.5 Gbit/s [28]. Using optical nonlinearity, IBM Corporation has realized all-optical switches with low bit error and a short switch time of only 2 ns [29].

4.6.2 Nonlinear Optical Devices

The optical intensity in the resonator is much larger than that in the bus waveguides. Thus, the nonlinear optical effects are enhanced, such as stimulated Raman scattering (SRS), two-photon absorption (TPA), and four-wave mixing (FWM). A centrosymmetric diamond lattice exhibits as first higher-order optical nonlinear effect the third-order susceptibility coefficients [30–32]. Second-order effects are possible with symmetry breaking structures (e.g., superlattices or strained layers).

The general Kerr effect is a change in the refractive index of a material in response to an applied electric field. John Kerr already discovered this effect in 1875 [33].The electro-optical Kerr effect is the case in which the electric field is due to the light itself. This causes a variation in the index of refraction, which is proportional to the local irradiance I of the light (I increases with the square of the electric field). The proportionality constant is called Kerr coefficient. The values of the Kerr coefficient are relatively small for most materials, on the order of $10-20$ m^2 W^{-1} for typical glasses. Therefore, beam intensities on the order of 1 GW cm^{-2} (such as those produced by lasers) are necessary to produce significant variations in the refractive index via the Kerr effect. The Kerr coefficient of Si at the telecom wavelength of 1.55 μm is roughly 100 times higher (Table 4.3) than that of fiber glass [34], but the unwanted absorption increases also due to TPA. TPA increases also with the beam intensity. By convention, the proportionality is written as a product of the wavelength in vacuum times the TPA coefficient. An often-used figure of merit (FOM) relates the Kerr coefficient with that product (wavelength*TPA coefficient). Si obtains only moderate FOM values of 0.3–0.4 at the telecom wavelength, but the FOM improves above a 2 μm wavelength because the TPA drops down at phonon energies below half of the bandgap. Unlike the situation with direct bandgap semiconductors such as GaAs, the Kerr coefficient in Si is always positive, which is assumed to be correlated with the indirect band character [35]. Many functionalities have already been demonstrated in silicon-based nonlinear devices, but often they suffer from reduced efficiency due to the rather strong TPA [36]. To overcome this limitation, alternative materials have been proposed by the scientific community, such as AlGaAs, chalcogenides, composite glasses, and nonlinear polymers in slot waveguides [34]. From the viewpoint of process integration, two materials emerged as specifically interesting, namely amorphous Si and single-crystalline SiGe [36] and graphene oxide [37].

Table 4.3 The nonlinear optical coefficients in silicon for a wavelength of 1.55 μm

	Measured
TPA coefficient	6 m/MW
Kerr coefficient	4.5×10^{-18} m^2/ W
SRS coefficient	95 m/MW

- **Raman laser**

The first silicon laser was realized through Raman effect, with a Fabry–Perot resonator. However, Fabry–Perot resonators need facet cutting, polishing, and antireflection coating, which seem to be complicated for fabrication and unsuitable for integrated photonics on chip.

Figure 4.32a shows a Raman laser based on a microring resonator. The excess free carriers are drawn out from the waveguides by a bias voltage, lowering the optical absorption. The threshold power is reduced by an order of magnitude using a microring resonator instead of a Fabry–Perot resonator. The pump efficiency and the output power both have been increased by several times with applied voltage [33], as seen in Fig. 4.32b. The Raman laser in silicon converts short wavelengths to longer wavelengths, broadening the wavelength range of the silicon laser. High-power pump lasing is done with an external laser. Hence, the application range is limited.

Applications of Microring Resonators | 87

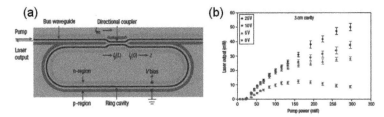

Figure 4.32 (a) The schematic image of a Raman laser and (b) output power versus pump power. Reprinted by permission from Springer Nature Customer Service Centre GmbH: Springer Nature, *Nature Photonics*, Ref. [38], copyright (2007).

- **All optical logic gates**

In terms of TPA and free carrier dispersion effect, one can realize all-optical logic gates, such as OR, AND, and NAND gates [32, 39, 40]. The microring resonator–based logic gates generally operate in this way: There are three input lights all at the resonant wavelengths: one is a signal light, and the others are control lights. The power of control lights is high, which causes optical nonlinearity and resonant peak shift of the microring resonators. If the control lights are switched, the status of the signal light will be changed, indicating optical logic operation.

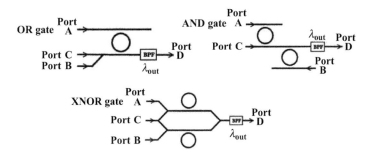

Figure 4.33 The schematics of OR, AND, and XNOR gates.

Three kinds of gates—OR, AND, and exclusive NOR (XNOR)—are presented in Fig. 4.33, where control lights are imported into A and B ports and the signal light is imported into the C port. For the OR gate, when both control lights are 0, the signal light is 0 at D, and as the control light is 1 at A or B, the resonant peaks are blue-shifted, making the signal light 1 at D. For the AND gate, just when both control lights are 1 at A and B, the two microring resonators are resonating and the signal light is nonresonant and becomes 1 at D. For the XNOR gate, when both A and B are 0 or 1, the two arms have no phase difference, so it is 1 at D; when A or B is 1, only one microring is resonant, the phase difference is π, enabling constructive interference, and so it is 0 at D.

4.6.3 Optical Buffer

In optical interconnect and optical computing, the optical buffer plays an important role since it is able to delay optical pulses and avoid signal blocking. Microring resonators serve as time delay lines and optical buffers. When the microring is resonating, the signal light will be captured and stay in the resonator for a moment. Time delay lines based on microring resonators have such configurations as all-pass filter (APF), coupled-resonator optical waveguide (CROW), and parallel-coupled double microrings with analogue to electromagnetically induced transparency (EIT) effect [19, 36], as shown in Fig. 4.34.

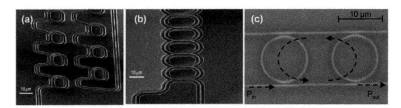

Figure 4.34 Configurations of (a) APF, (b) CROW, and (c) double microrings with analogue to EIT.

It is clear that the time delay increases with the number of cascaded microrings. However, the insertion loss increases, too. Increasing the Q factor will enlarge the delay time but decrease the bandwidth. Therefore, one should find a compromise between time delay, bandwidth, and insertion loss in the design. IBM Corporation realized a time delay of 510 ps and a 10-bit buffer, with series-coupled microring resonators with a footprint of only 0.09 mm^2. In the parallel-coupled double microrings, each ring acts as a "reflector." Thus, the two reflectors construct a "Fabry–Perot resonator" with analogue to EIT. Table 4.4 presents some results of optical buffers based on silicon microring resonators.

Table 4.4   Optical buffers based on silicon microring resonators

	Time delay (ps)	Area (mm²)	Loss (dB)	Storage (bits)
APF of 56 rings	510	0.09	22	10 at 20 bps
CROW of 100 rings	220	0.045	23	1 at 5 bps
EIT-analogue rings	25	3×10^{-4}	4	—
Waveguides of 4 cm	500	1.13	7	>>10

4.7 Summary

Several decades have passed since the invention of microring resonators. Microring resonators have made great progress in the first decade of the 21st century because of their broad use in silicon photonics. Unlike facet mirrors and distributed Bragg reflectors, they are easily integrated in planar Si microelectronics technology and they are very versatile combined with other components, such as waveguides, mirrors, and MZIs.

There is constant progress in microring resonator design and fabrication, especially in terms of:

- Improvement in the processes to develop the performance. Industrial DUV lithography and nanoimprint lithography are employed and optimized.
- Development of novel functional photonic devices such as optical filters, modulators, switches, lasers, and logic gates.
- Application as reconfigurable wavelength routers to increase transmission bandwidth in the optical networks mainly based on point-to-point communications [37].
- Paving the way to new application fields. For example, on-chip sensors and biosensors became a strong research focus recently.

However, some inherent challenges confront microring technology and system performance, such as low processing tolerance, temperature sensitivity, and polarization dependence. Promising solution paths include (i) employing advanced techniques to increase the fabrication accuracy and optimizing the design of devices to improve the processing tolerance, (ii) introducing nanoelectronic technology and employing temperature-maintained modules to keep a stable microring resonator working, and (iii) adding polarization converters and polarization splitters to balance the two polarizations and achieve polarization independence in system.

In summary, a microring resonator is a versatile component in integrated photonics that serves as a unit in various functional systems and plays an important role in optical communication, optical interconnect, and optical computing. With the progress in research and technology, microring resonators have become more an industrial component and find even more applications [38].

Acknowledgments

We thank Qingzhong Huang and Xi Xiao for the collection of basic information and figure drawings.

References

1. E. A. J. Marcatilli (1969). Bends in optical dielectric waveguides, *Bell Syst. Tech. J.*, **48**, 2103–2132.

2. B. E. Little, S. T. Chu, H. A. Haus, et al. (1997). Microring resonator channel dropping filters, *J. Lightwave Technol.*, **15**(6), 998–1005.

3. A. Yariv (2000). Universal relations for coupling of optical power between microresonators and dielectric waveguides, *Electron. Lett.*, **36**(4), 321–322.

4. L. Caruso and I. Montrosset (2003). Analysis of a racetrack microring resonator with MMI coupler, *J. Lightwave Technol.*, **21**(1), 206–210.

5. J. K. S. Poon, J. Scheuer, S. Mookherjea, et al. (2004). Matrix analysis of microring coupled-resonator optical waveguides, *Opt. Express*, **12**(1), 90–103.

6. S. Darmawan, Y. M. Landobasa and M. K. Chin (2007). Pole-zero dynamics of high-order ring resonator filters, *J. Lightwave Technol.*, **25**(6), 1568–1575.

7. R. Grover, P. P. Absil, T. A. Ibrahim, et al. (2004). III-V semiconductor optical micro-ring resonators, *Microresonators as Building Blocks for VLSI Photonics: International School of Quantum Electronics*, *AIP Conference Proceedings*, Vol. 709, 110–29 2004

8. Y. Yanagase, S. Suzuki, Y. Kokubun, et al. (2002). Box-like filter response and expansion of FSR by a vertically triple coupled microring resonator filter, *J. Lightwave Technol.*, **20**(8), 1525–1529.

9. J. Heebner, R. Grover and T. Ibrahiml (2007). *Optical Microresonators: Theory, Fabrication, and Applications* (Springer-Verlag, Berlin).

10. Q. Z. Huang, Y. D. Yu and J. Z. Yu (2009). Experimental investigation on submicron rib waveguide microring/racetrack resonators in silicon-on-insulator, *Opt. Commun.*, **282**, 22–26.

11. T. Barwicz, M. R. Watts, M. A. Popovic, et al. (2007). Polarization-transparent microphotonic devices in the strong confinement limit, *Nat. Photonics*, **1**(1), 57–60.

12. X. Yan, C. S. Ma, Y. Z. Xu, et al. (2005). Characteristics of vertical bent coupling between straight and curved rectangular optical waveguides, *Optik*, **116**(8), 397–403.

13. D. Marcuse (1972). *Light Transmission Optics*, 2nd ed. (Van Nostrand Reinhold, New York).

14. V. Subramaniam, G. N. DeBrabander, D. H. Naghski, et al. (1997). Measurement of mode field profiles and bending and transition losses in curved optical channel waveguides, *J. Lightwave Technol.*, **15**(6), 990–997.

15. K. K. Lee, D. R. Lim, H. C. Luan, et al. (2000). Effect of size and roughness on light transmission in a $Si/SiO2$ waveguide: experiments and model, *Appl. Phys. Lett.*, **77**(11), 1617–1619.

16. P. K. Tien (1971). Light waves in thin films and integrated optics, *Appl. Opt.*, **10**(11), 2395.

17. S. M. Zheng, H. Chen and A. W. Poon (2006). Microring-resonator cross-connect filters-in silicon nitride: rib waveguide dimensions dependence, *J. Sel. Topics Quantum Electron.*, **12**(6), 1380–1387.

18. P. Dumon, W. Bogaerts, V. Wiaux, et al. (2004). Low-loss SOI photonic wires and ring resonators fabricated with deep UV lithography, *IEEE Photonics Technol. Lett.*, **16**(5), 1328–1330.

19. F. N. Xia, L. Sekaric and Y, Vlasov (2007). Ultracompact optical buffers on a silicon chip, *Nat. Photonics*, **1**(1), 65–71.

20. Q. F. Xu, B. Schmidt, S. Pradhan, et al. (2005). Micrometre-scale silicon electro-optic modulator, *Nature*, **435**(7040), 325–327.

21. S. J. Xiao, M. H. Khan, H. Shen, et al. (2007). Compact silicon microring resonators with ultra-low propagation loss in the C band, *Opt. Express*, **15**, 14467–14475.

22. U. Plachetka, N. Koo, T. Wahlbrink, et al. (2008). Fabrication of photonic ring resonator device in silicon waveguide technology using soft UV-nanoimprint lithography, *IEEE Photonics Technol. Lett.*, **20**(5–8), 490–492.

23. Q. Z. Huang, J. Z. Yu, S. W. Chen, et al. (2008). Design, fabrication and characterization of a high-performance microring resonator in silicon-on-insulator, *Chinese Phys. B*, **17**(7), 2562–2566.

24. F. N. Xia, L. Sekaric and Y. A. Vlasov (2006). Mode conversion losses in silicon-on-insulator photonic wire based racetrack resonators, *Opt. Express*, **14**(9), 3872–3886.

25. Q. Z. Huang, Y. D. Yu and J. Z. Yu (2009). Design and realization of a microracetrack resonator based polarization splitter in silicon-on-insulator, *J. Opt. A: Pure Appl. Opt.*, **11**, 015506.

26. F. N. Xia, M. Rooks, L. Sekaric, et al. (2007). Ultra-compact high order ring resonator filters using submicron silicon photonic wires for on-chip optical interconnects, *Opt. Express*, **15**, 11934–11941.

27. S. Xiao, M. H. Khan, H. Shen, et al. (2007). Multiple-channel silicon micro-resonator based filters for WDM applications, *Opt. Express*, **15**(12), 7489–7498.

28. Q. F. Xu, S. Manipatruni, B. Schmidt, et al. (2007). 12.5 Gbit/s carrier-injection-based silicon micro-ring silicon modulators, *Opt. Express*, **15**(2), 430–436.

29. Y. Vlasov, W. M. J. Green and F. Xia (2008). High-throughput silicon nanophotonic wavelength-insensitive switch for on-chip optical networks, *Nat. Photonics*, **2**(4), 242–246.

30. O. Lin, O. J. Painter and G. P. Agarwal (2007). Nonlinear optical phenomena in silicon waveguides: modeling and applications, *Opt. Express*, **15**, 16606.

31. L. Zhang, A. M. Agarwal, L. C. Kimerling and J. Michel (2014). Nonlinear group IV photonics based on silicon and germanium: from near-infrared to mid-infrared, *Nanophotonics*, **3**, 247–268.

32. H. K. Tsang and Y. Liu (2008). Nonlinear optical properties of silicon waveguides, *Semicond. Sci. Technol.*, **23**(6), 064007.

33. J. Kerr (1875). A new relation between electricity and light:Dielectric media birefringent, *Phil. Mag.*, **50**(332), 337–348.

34. J. Leuthold, C. Koos and W. Freude (2010). Nonlinear silicon photonics, *Nat. Photon.*, **4**, 535–544.

35. M. Dinu, F. Quochi and H. Garcia (2003). Third-order nonlinearities in silicon at telecom wavelengths, *Appl. Phys. Lett.*, **82**, 2954–2956.

36. C. Lacava, M. A. Ettabib and P. Petropoulos (2017). Nonlinear silicon photonic signal processing devices for future optical networks, *Appl. Sci.*, **7**, 103.

37. X. Xu et. al. (2017). Observation of third-order nonlinearities in graphene oxide film at telecommunication wavelength, *Sci. Rep.*, **7**, 9646.

38. H. S. Rong, S. B. Xu, Y. H. Kuo, et al. (2007). Low-threshold continuous-wave Raman silicon laser, *Nat. Photonics*, **1**(4), 232–237.

39. T. A. Ibrahim, K. Amarnath, L. C. Kuo, et al. (2004). Photonic logic NOR gate based on two symmetric microring resonators, *Opt. Lett.*, **29**(23), 2779–2781.

40. T. A. Ibrahim, R. Grover, L. C. Kuo, et al. (2003). All-optical AND/NAND logic gates using semiconductor microresonators, *IEEE Photonics Technol. Lett.*, **15**(10), 1422–1424.

41. Q. F. Xu, J. Shakya and M. Lipson (2006). Direct measurement of tunable optical delays on chip analogue to electromagnetically induced transparency, *Opt. Express*, **14**(14), 6463–6468.

42. J. Thomson, F. Y. Gardens, J.-M. Fédéli, S. Zlatanovic, Y. Hu, B. P. P. Kuo, E. Myslivets, N. Alic, S. Radic, G. Z. Mashanovich and G. Reed (2012). 50-Gb/s silicon optical modulator, *IEEE Photon. Technol. Lett.*, **24**(4), 234–236.

43. M. Calvo et. al. (2019). Ring resonator designed for biosensing applications manufactured on 300 mm SOI in an industrial environment, *Jpn. J. Appl. Phys.*, **58**, SBBE02.

Chapter 5

Optical Couplers

5.1 Introduction

Optoelectronics integrated circuits (OEICs) have progressed rapidly since the 1960s. For scaling down in integrated circuits, the waveguide structure in the OEIC requires a high refractive index contrast between waveguide layer and cladding layer. Among all the candidate platforms for waveguide structure applied in communication wavelengths, silicon-on-insulator (SOI) has been extensively investigated because of its good optical properties and compatibility with complementary metal-oxide-semiconductor (CMOS) fabrication processes. First, the absorption loss of SOI is relatively low in the infrared region and this is a critical property for optical communication and optical interconnect on chips. Secondly, the high refractive index contrast ($\Delta n \approx 2$) between the silicon core and silica cladding layers supports extreme light confinement and allows the waveguide core to be shrunk down to a submicron cross section. Besides, SOI is compatible with the microelectronics fabrication process, so low-cost monolithic integration of optical and electrical circuits is possible. All these properties ensure that SOI is a promising platform for OEIC [1]. These silicon circuits are also named photonic integrated circuits (PICs) due to the dense packing of optical functional devices and interconnects. Submicron

---

Silicon-Based Photonics
Erich Kasper and Jinzhong Yu
Copyright © 2020 Jenny Stanford Publishing Pte. Ltd.
ISBN 978-981-4303-24-8 (Hardcover), 978-981-4303-25-5 (eBook)
www.jennystanford.com

waveguides can be utilized in many compact structures with outstanding performance [2, 3]. However, there are two challenges for its large-scale application in optical communication and computing systems:

- Single-mode waveguide is polarization sensitive. Fortunately, polarization independence can be obtained by deep-etched submicron rib waveguides [3].
- Optical coupling between an optical fiber and a nanophotonic waveguide causes a high insertion loss.

An essential question is how to couple light efficiently from the optical fiber to the PICs. As the SOI platform offers a high refractive index difference between core and cladding, the cross section of single-mode waveguides is a factor of $\sim10^{-3}$ smaller than that of standard single-mode fibers and hence the large dimension mismatch induces large coupling losses. To bridge the gap between silicon PICs and the outside world, highly efficient coupling structures have to be developed, with the stringent conditions that they should be compact in size, broadband, and cost effective in fabrication.

The coupling loss between an optical fiber and a nanophotonic waveguide mainly consists of five parts: (i) lateral displacement loss, (ii) longitudinal transmission loss, (iii) axial tilting loss, (iv) modes mismatch loss, and (v) numerical-apertures mismatch loss. The preceding three loss mechanisms are caused by an alignment error, while the last two losses do not depend on the accuracy of alignment between the optical fiber and the nanophotonic waveguide. Typical optical fibers have mode sizes of the order of approximately 10 μm, contrasting greatly with the submicron size of Si-based submicron waveguides. The mismatch in both mode size and effective refractive index will result in radiation mode and back reflection from interfaces, which cause a high insertion loss for light coupling. Therefore, efficient optical coupling between an optical fiber and a nanophotonic waveguide is a key challenge for OEIC.

Several approaches for efficient light coupling have been proposed and experimentally demonstrated, the most popular approaches being tapered spot-size converters, prism couplers, and grating couplers. Table 5.1 lists different types of silicon-based optical couplers, including their coupling efficiency, bandwidths, and advantages and disadvantages.

Introduction | 97

Table 5.1 Comparison of different types of silicon couplers

	Coupling efficiency	Bandwidth	Advantage	Disadvantage
Tapered spot-size converters	Low	Very broad	Easy to fabricate	Large dimension, hard to integrate with a submicron waveguide, and of low coupling efficiency
Inverted tapered spot-size converters	Highest (90%)	Very broad	High alignment tolerance	Hard to fabricate and polarization sensitive
Inverted prism couplers	High (50%)	Broad	Easy to fabricate, flexible application, and reliable alignment	Hard to integrate, polarization sensitive, and extra epoxy adhesive
Graded index (GRIN) waveguide couplers	Very high (80%)	Broad	High fabrication tolerance	Hard to fabricate and integrate, low alignment tolerance, and high polarization sensitivity
Grating couplers	Very high	Broad	Easy to fabricate and integrate, high alignment tolerance, and no facet polishing	Polarization sensitive

A tapered spot-size converter has a very broad bandwidth but low coupling efficiency and a relatively large size, and it increases complexity for integration with a submicron waveguide. An inverted tapered spot-size converter offers the highest coupling efficiency—as much as 89%—but it is polarization sensitive and has stringent alignment tolerances and complex fabrication processes. A graded index (GRIN) waveguide coupler has high efficiency, of 78%, but it has stringent alignment tolerances and complex fabrication processes; it is also polarization sensitive. A grating coupler is relatively easy to fabricate and integrate. It has a high alignment tolerance and needs no facet polishing.

Inverted tapered spot-size converters [4] reported by IBM offered a very impressive coupling loss of only 0.5 dB per interface. Grating couplers [5] developed by Ghent University have solved the problem of the big mismatch between optical fiber and nanophotonic waveguide in dimension and realized vertical light coupling.

Light couplers are essential components of both optical interconnects and silicon photonics systems. In this chapter, working principles and properties of different silicon couplers are introduced and compared with each other.

5.2 Spot-Size Converter

Tapered structure waveguides, with dimension and symmetry differences, are widely used in spot-size converters for light coupling, which can significantly reduce the mismatch caused by the refractive index. The coupling efficiency of tapered spot-size converters depends on mode match between optical fiber and waveguide, surface refraction, and roughness. The mode mismatch can especially introduce a high insertion loss.

Tapered structure waveguides can convert the large mode spot-size of an optical fiber into a small mode spot-size so that the light can be coupled better into a nanophotonic waveguide, or vice versa. Typically, there are two kinds of spot-size converters: tapered and inverted. Not only the mode size but also the mode shape can be changed. This helps realize the perfect match between fiber optical mode and waveguide optical mode and improve light coupling efficiency accordingly. Tapered spot-size converters are intuitively the most obvious structures as their width changes gradually.

This allows a match between an optical fiber and a nanophotonic waveguide. Usually, tapered spot-size converters can be classified as lateral [6, 7], vertical [8–14], and combined (tapered both vertically and laterally) tapers [15, 16]. It has been demonstrated that inverted tapered [17–19] and slot-waveguide structures [20] are also spot-size converters.

5.2.1 Tapered Spot-Size Converters

On Optical Fiber Communication Conference 2003, Bookham's group first reported a vertical tapered spot-size converter for light coupling between optical fibers and SOI electronically variable optical attenuators; its coupling loss was only 0.5 dB per interface [6]. On the basis of Bookham's research, the Institute of Semiconductors, Chinese Academy of Sciences, made some improvements on its square rib structure and divided it into an upper rib and a middle rib, as shown in Fig. 5.1 [7].

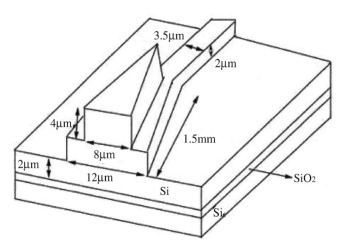

Figure 5.1 Tapered spot-size converter. © 2005 IEEE. Reprinted, with permission, from Ref. [7].

As a tapered spot-size converter, its end widths should be close to an optical fiber and a nano-optical waveguide, to avoid optical mode mismatch, which can cause a lot of insertion loss on taper connection between fibers and waveguides. When the light couples into the tapered spot-size converter, it first scatters across the three layers of the waveguide structure, which is as thick as 8 μm. Then it gradually focuses on the middle rib and the planar waveguide. Finally, it steadily propagates in the single-mode rib waveguide and the mode converting process ends in this tapered spot-size converter.

The insertion loss of the spot-size converters depends on their length and end width. The longer the spot-size converter, the lesser is the insertion loss and the slower is the optical mode converting process. The wide end of the converter will introduce a high additive loss by the light leakage. Simulation analyses have shown that the insertion loss, including the transmission loss in tapers, the coupling loss with the fiber, and the refraction loss at the interface, was 1.81 dB, in which coupling loss with the fiber was only 0.44 dB. Compared with direct light coupling from an optical fiber to a single-mode nano-optical waveguide, where the insertion loss is approximately 9 dB when alignment condition is the best, a tapered spot-size converter can reduce 7 dB of the insertion loss between an optical fiber and a nano-optical waveguide.

This kind of spot-size converter has shown very high coupling efficiency. However, there are still some problems, which limit its application. To avoid light leakage, the tapered spot-size converter is designed to be so long (1500 μm) that monolithic integration becomes unattractive. The upper rib is 4 μm high and easy to break during the fabrication and test processes. To reduce reflection loss of this spot-size converter, an antireflection coating is necessary for the light input and output port. Furthermore, polarization sensitivity limits its application, too.

Figure 5.2 shows a 3D adiabatically tapered structure [8], which can efficiently couple light from an optical fiber or free space to a chip. This structure of optical waveguides was fabricated directly on an SOI wafer. Fabrication processes included (i) writing a single grayscale mask by using a high-energy electron beam on a high-energy-beam-sensitive glass, (ii) ultraviolet grayscale lithography, and (iii) inductively coupled plasma etching. The input and output facet dimensions were 10 μm, approximately equal to the core diameter of a single-mode fiber. Its coupling efficiency increased with taper length for a given central waveguide as a result of the gradual mode conversion. The cross-sectional variation of the tapers was limited to vertical tapering because of the resolution constraints of the thick resist, so it did not achieve horizontal mode conversion. However, this did not limit its applicability. The tapering geometry was fixed to linear tapers to simplify fabrication, but the developed process could be easily extended to other geometries, such as sinusoidal or quadratic tapers. It should be pointed out that the coupling efficiency can be high by optimizing geometry if the taper is sufficiently long (>600 μm).

Figure 5.2 3D adiabatically tapered spot-size converter. Reprinted with permission from Ref. [8] © The Optical Society.

An ideal structure of the spot-size converter is the one that adiabatically tapers in both vertical and lateral directions; it can also be monolithically integrated with the OEIC. However, such structures cannot be achieved with standard lithography techniques. Confluent Photonics Inc. has fabricated such structures by gray tone lithography [15] and polishing [16], respectively. However, gray tone lithography introduces high losses due to misalignment and the polishing process could not be used in integrated fabrication of OEIC.

To fabricate vertically tapered structures, the dip-etch process [9], the dynamic etch mask technique [10], diffusion limited etch [11], and stepped etching [12] have been reported. All these methods have disadvantages of either low reproducibility or labor-intensive processing. Other techniques, like dry etching by a shadow mask [13], have been used in fabricating vertical tapers, but they do not allow processing on the wafer scale. Fabrication processes of the tapered spot-size converter of Fig. 5.2 are rather complex, but this converter is well suited for dense wafer-scale integration, so it can be further processed to realize other OEIC components. Its measured coupling efficiency of 45%–60% was lower than the theoretically expected value of 82%, mainly resulting from the scattering losses caused by surface roughness. As other tapered spot-size converters, antireflection coatings were used to prevent

102 | *Optical Couplers*

back reflection from the facets. This added complexity and cost. Furthermore, the minimization of the taper structure could cause serious surface roughness, resulting in higher back reflection and insertion loss.

5.2.2 Inverted Tapered Spot-Size Converters

According to single-mode condition, a strong light confinement effect can be achieved in a typical SOI single-mode waveguide where the mode profile is concentrated in the waveguide core and its effective refractive index approximately equals that of the silicon. However, the light confinement effect tends to be weak with the shrinking waveguide dimensions. When the waveguide is less than 150 nm wide, this confinement is very limited. An inverted tapered spot-size converter consists of a waveguide laterally tapered to a nanometer-sized tip at the facet in contact with the fiber. At the tip the mode field profile becomes delocalized from the waveguide core. And this delocalization of the mode field profile increases the mode overlap with the optical fiber mode. In addition, most of the mode field resides in the SiO_2 cladding region at the tip, causing the effective index to be close to that of the fiber, which results in negligible back reflections and high coupling efficiency.

A research group of Cornell University proposed and demonstrated an inverted tapered spot-size converter as shown in Fig. 5.3 [17]. It was a nano-optical waveguide laterally tapered to a nanometer-sized tip at the facet in contact with the fiber. The nanotaper and waveguides were fabricated on an SOI wafer with a SiO_2 layer as an optical buffer. SOI provides high-index difference and permits compatibility with integrated electronic circuits. The power overlap and the mode mismatch loss depend on the tip width for both transverse electric (TE)- and transverse magnetic (TM)-like modes. When the tip is 50 nm wide, the mode mismatch loss is only 0.4 dB. To convert the low-confined local mode at the nanotaper tip into the high-confined waveguide mode, a short tapered transition was employed by gradually varying both sidewalls in a symmetric parabolic transition toward the final waveguide width, where the parabola vertex is located at the nanotaper tip. Therefore, the taper length, which is necessary to convert the mode, also depends on the width of the nanotaper tip.

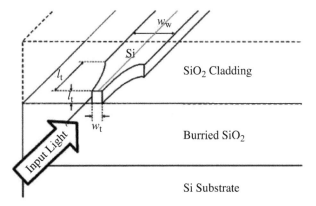

Figure 5.3 Inverted tapered spot-size converter by Cornell's group. Reprinted with permission from Ref. [17] © The Optical Society.

This kind of inverted tapered spot-size converter had a well-considered structure. However, its insertion loss was still higher than 3 dB. Furthermore, it was very sensitive to the polarization state. Simulation showed that the insertion losses for TM-like and TE-like modes at 1550 nm were 3.3 dB and 6 dB, respectively. The total measured insertion loss was 9.2 dB, including about 5 dB loss caused by the nano-optical waveguides, and this was quite close to the simulated result. The insertion loss originated from the mode mismatch loss between the optical fiber and tip facet modes and from mode conversion of the low-confined mode at the tip facet into the high-confined mode in the waveguide. Meanwhile, its misalignment tolerance was relatively large and the additional insertion loss for 1.2 μm misalignment in both horizontal and vertical directions was only 1 dB. This mainly resulted from the large field dimensions at the tip.

Several inverted tapered spot-size converters with good properties have been proposed and demonstrated. An inverted tapered spot-size converter as shown in Fig. 5.4 had only 1 dB measured mode conversion loss and 3.5 dB total insertion loss [18]. Though the insertion loss is relatively high, the converter has the potential to be used as a building block for OEIC if the random scattering loss caused by the sidewall in the silicon waveguide can be reduced. An inverted taper of similar structure with a very impressive coupling loss of only 0.5 dB was reported [19]. Among the reported SOI tapered or inverted tapered spot-size converters, different coupler solutions optimize either small footprint or high coupling efficiency or simple fabrication processes. However, the insertion efficiency is polarization sensitive.

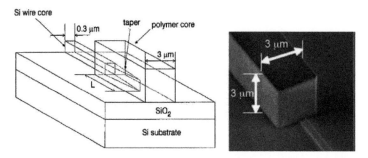

Figure 5.4 Schematic diagram of an inverted tapered spot-size converter. Republished with permission of Institution of Engineering and Technology (IET), from Ref. [18]. Copyright (2002); permission conveyed through Copyright Clearance Center, Inc.

5.2.3 Slot-Waveguide Spot-Size Converter

The slot-waveguide spot-size converter shown in Fig. 5.5 realizes a novel optical fiber-to-waveguide coupler concept [20]. It consists of an SOI substrate, slot waveguides, and cladding layers.

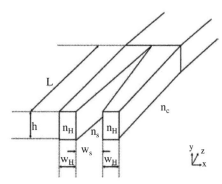

Figure 5.5 Schematic of the slot-waveguide spot-size converter. Reprinted with permission from Ref. [20] © The Optical Society.

The effective refractive index of this slot-waveguide structure is close to that of the standard single-mode fiber. The gap of the slot was tapered to be less than 150 nm at the facet in contact with the fiber where the mode field size increased because of the mode delocalization. The overlap of the mode fields of waveguide and fiber was enhanced, and optical low loss coupling from a fiber to the slot waveguide could be achieved accordingly.

To satisfy the continuity of the normal component of electric flux density, Maxwell's equations state that the corresponding electric field must undergo a large discontinuity with a much higher amplitude on the low-index side for a high-index-contrast interface. So discontinuity in the slot waveguide can be used to strongly enhance and confine light in a nanometer-wide region of a low-index material. The light fields overlap on the low-index side and localize to both slot waveguides finally. Therefore, high-efficiency light coupling is obtained.

The mode conversion loss of the slot-waveguide spot-size converter is relatively low. Simulated mode conversion loss with the air taper was less than 0.5 dB for a TE-like mode when the length of this converter was less than 90 nm. Furthermore, the mode conversion loss was negligible for the TM-like mode. The theoretical insertion loss was 1.8 dB, which mainly originates from mode mismatch at the facet.

The Institute of Microelectronics of Singapore presented a slot-waveguide spot-size converter for light coupling between a fiber and a submicron silicon nitride waveguide [21]. The coupling efficiency of the fabricated double-tip coupler was improved by as much as over 2 dB per coupling facet. This slot-waveguide spot-size converter has small dimensions but high coupling efficiency; its fabrication scheme is compatible with CMOS processes. However, the complexity of the fabrication processes, high requirements for the photoresist of electron beam lithography, and polarization sensitivity limit its widespread use.

5.3 Prism Coupler

5.3.1 Inverted Prism Coupler

Typically, a planar waveguide consists of three layers: cladding, guiding, and substrate. The imposition of the electromagnetic boundary condition leads to two physical conditions: total internal reflection in the guiding layer and a phase matching condition in each layer, which results in a set of discrete modes and their corresponding mode angles. With the prism coupler, guided waves with the appropriate mode angle can be introduced and high coupling efficiency can be achieved.

Figure 5.6 shows the schematic of inverted prism couplers [22]. The coupler incorporates the advantages of the vertically tapered spot-size converter and prism couplers and offers the flexibility for planar integration.

Polishing grayscale lithography or inductively coupled plasma (ICP) etching could be used to fabricate such kinds of structures on SOI substrates. The theoretical coupling efficiency is as high as 77%, and the measured coupling efficiency was 46% when it was coupled with a nanophotonic waveguide 0.25 m in diameter. The inverted prism had the advantages of flexible application, simple alignment process, high coupling efficiency, and broad 3 dB bandwidth (80 nm). However, an epoxy adhesive was needed for bonding with a nanophotonic waveguide. Besides, the inverted prism coupler was polarization sensitive and close to the waveguide, which easily disturbs the waveguide structures.

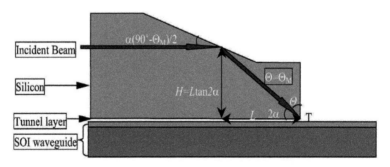

Figure 5.6 Schematic of an inverted prism coupler. Reprinted with permission from Ref. [22] © The Optical Society.

5.3.2 Graded Index Half-Prism Coupler

Figure 5.7 shows a GRIN half-prism coupler based on SOI. Its operation principal is similar to that of a conventional cylindrical

GRIN lens. Input light periodically converges in the SOI waveguide core and then diverges. The refractive index of an ideal planar GRIN waveguide lens decreases with a quadratic dependence on distance from the waveguide surface [23]. However, much simpler GRIN index profiles consisting of two or three uniform layers with a stepwise decreasing index profile perform almost as effectively as the ideal structure. Even the simplest case of a single uniform layer of a slightly lower index than the waveguide core could increase the optical power coupled into a small SOI waveguide from a fiber or a large input beam by several times.

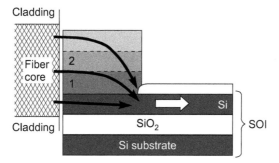

Figure 5.7 Schematic of the monolithic GRIN coupler structure.

Fabricating a GRIN coupler involved two key steps. The first was to deposit and pattern the a-Si GRIN layers on the SOI waveguide to form the coupler. The second was to create a lithographically defined input facet so that the correct coupler length was produced. Ridges were etched by reactive ion etching. The waveguide coupler sections were defined by etching a window in this SiO_2 layer over the waveguide adjacent to the eventual input facet position. The refractive index over the die area showed some spatial variation, ranging from 3.36 to 3.40. This index range is less than the index of Si, as required for effective GRIN lens focusing. The ICP etch process was developed to fabricate the input and output facets. To couple with a 0.5 μm silicon waveguide, the theoretical coupling efficiency was 71% and it could be increased to 78% by optimizing operation from a 4 μm thicker GRIN lens. The experimental and theoretical coupling efficiencies were in good qualitative agreement.

108 | *Optical Couplers*

Besides the high light coupling efficiency, there are many other advantages of the GRIN coupler. The precision requirement for fabrication is only 1 μm, and a simple optical lithography process is qualified. Compared with the inversed tapered spot-size converter, the GRIN coupler does not need such an accurately thick cladding layer. The transmission loss in this coupler is negligible as its focal length is about 10 μm. However, the single layer GRIN coupler is sensitive to the mode shape of the input light and to the lateral misalignment of the fiber.

5.4　Grating Coupler

A grating coupler is a sort of novel solution to light coupling from an optical fiber to a nano-waveguide (and vice versa). Typically, there are both vertical [24–30] and horizontal [31–34] structures for light coupling. Grating couplers have a relatively high theoretical coupling efficiency. Nevertheless, the conventional grating couplers [32, 33] for OEIC have to overcome difficulties that stem from mode matching, the complexity of fabrication, the sensitivity to wavelengths, and the polarization state. Furthermore, the stringent light incident angle requirement needs special housings for packaged circuits or precise optical alignment for on-wafer measurements. However, the grating couplers have been investigated widely and improved significantly.

In the grating coupler with a horizontal coupling structure, the light from a fiber is directly coupled into a waveguide of a large cross section, then guided to a grating structure etched on this large waveguide, and finally coupled into the small-dimension waveguide by diffraction. In the vertical coupling structure, a 2D spot-size converter is needed for the light coupling between a nanophotonic waveguide and a large cross-section waveguide in which the light is coupled from a fiber by the grating structure.

Grating couplers offer the following advantages:

- There is no need for additional wafer cleaning and facet polishing, which prevents not only complex post processes but also the potential threat to the circuits.
- Alignment tolerances are relatively high because the coupling area of the grating is similar to the cross section (10 μm diameter) of an optical fiber.

- They enable wafer-scale tests of PICs because of the flexibility of light coupling in/out from OEIC.
- Fabrication is compatible with CMOS processes.

The grating is formed in a wide waveguide section by linear grooves with periodic submicron spacing. The fabrication of the grooves utilizes techniques and processes applied in microelectronics manufacture, lithography, and etching for pattern formation.

5.4.1 Grating Coupler with a Vertical Coupling Structure

As mentioned above, the grating couplers with a vertical coupling structure should work with the spot-size converter. The light from a fiber is coupled into the waveguide of a large cross section by a grating structure, and then the spot-size converter can realize the mode match between the nanophotonic waveguide and the large cross-section waveguide. The fiber is aligned nearly perpendicular to the waveguide surface, and light coupling between an optical fiber and a nanophotonic waveguide is achieved.

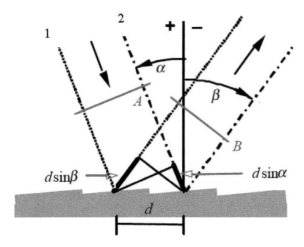

Figure 5.8 Illustration of a diffraction grating.

As shown in Fig. 5.8, parallel rays of monochromatic radiation from a single beam in the form of ray 1 and ray 2 are incident on a diffraction grating at angle α. These rays are then diffracted at angle

β. If the total path difference between ray 1 and ray 2 is equal to an integer multiple of the wavelength, constructive interference occurs. That is,

$$d(\sin \alpha + \sin \beta) = m\lambda, \quad (5.1)$$

where d is the grating period, the integer m is the diffraction order and $\lambda = \lambda_0/n_{eff}$, λ_0 is the vacuum wavelength, and n_{eff} is the effective refractive index of the dielectric. Equation 5.1 is the general grating equation.

When the incident illumination is perpendicular to the grating surface, that is, the incident angle $\alpha = 0$ and the diffracted angle $\beta = 90$, the grating formulation of Eq. 5.1 reduces to

$$d\, n_{eff} = m\lambda_0, \quad (5.2)$$

that is the diffraction grating equation for normal incidence. After incidence of light on a proper designed grating, the diffracted light is parallel to the grating surface direction and thereby can couple into the parallel nanophotonic waveguide. This is how the grating couplers of a vertical coupling structure work.

Equation 5.1 describes the relationship between the grating period, the effective refractive index of the dielectric, and the operating center wavelength. All these should be carefully considered for grating design. For convenience, only the first-order diffraction ($m = 1$), which contributes most to the light coupling, is considered.

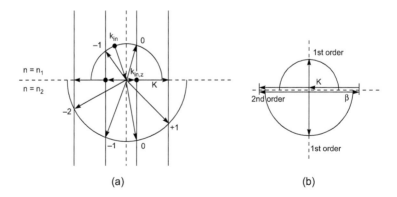

Figure 5.9 (a) A wave vector diagram for Bragg condition diffraction and (b) a wave vector diagram for second-order diffraction.

The Bragg condition is the most fundamental formula concerning periodic structures. As is shown in Fig. 5.9a, the Bragg condition describes the relation between the wave vectors of the incident and diffracted waves. If a grating consists of two materials with different refractive indexes and has the same period along the z axis, the Bragg equation is

$$k_z = k_{in,z} + mK, \qquad (5.3)$$

where $k_{in,z}$ is the z component of $k_0 n_1$ in material 1, k_z is the z component of $k_0 n_2$ in material 2, and $K = \dfrac{2\pi}{d}$. This equation suits all kinds of grating structures, but for a waveguide grating coupler, a similar formula, in which the incident wave is replaced by the guided mode of the waveguide, can be used. It is:

$$k_{in,z} = \beta - mK, \qquad (5.4)$$

where $\beta = \dfrac{2\pi}{\lambda_0} n_{eff}$. Equation 5.4 agrees with Eq. 5.2 for vertical incidence because then the in-plane component of the incoming wave $k_{in,z}$ is 0. But Eq. 5.4 also describes the coupling of light from an oblique alignment of the fiber with a finite component $k_{in,z}$. For a given grating period length d, the wavelength of the diffracted light can be changed by varying the inclination angle. The output of light from the waveguide to the fiber obeys the same laws. Figure 5.9b shows the wave vector diagram of a second-order grating for a waveguide grating coupler with a parallel guided wave. In this case, the first-order diffraction is vertically coupling both up and down out of the waveguide. Therefore, if light from the waveguide is diffracted by the grating structure etched on the waveguide, the diffraction wave up out of the waveguide can be coupled into an optical fiber.

The grating couplers usually consist of a grating structure of hundreds of micrometers. The coupling efficiency can be as high as 70% [31]. However, this is only the power that is coupled out of the waveguide, and not the coupling efficiency to fiber. When such a long grating structure is used, a curved focusing grating can be used to couple to a fiber. Though the output coupling efficiency is relatively high, the bandwidth of this kind of long grating coupler is narrow, which is caused by the small coupling strength.

The group from Ghent first reported an out-of-plane grating coupler for vertical coupling between 240-nm-thick GaAs–AlO$_x$ waveguides and single-mode fibers [24]. In 2007, they proposed and demonstrated a grating coupler of only 10 µm length in SOI, as shown in Fig. 5.10 [25]. This SOI grating coupler had a Si core layer of 220 nm and a buried oxide layer of 1 µm. The etch depth was 70 nm, and the grating period was 630 nm. After grating diffraction, besides reflected and transmitted wave, there were upward- and downward-propagating waves. Since the coupling strength resulting from a deep etched groove is large enough, the light power was mostly contained in the upward- and downward-propagating waves. Thereby, the coupling efficiency mainly depends on the light power contained by the upward-propagating wave. The grating directionality is defined as the ratio between the optical power diffracted toward the fiber and the total diffracted optical power. High grating directionality is important to achieve a high coupling efficiency.

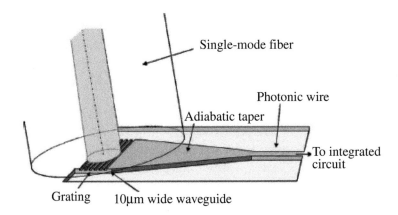

Figure 5.10 Schematic of Ghent's grating coupler. © 2005 IEEE. Reprinted, with permission, from Ref. [25].

The thickness of the buried oxide layer influences the efficiency of SOI grating couplers as the downward-propagating wave partially reflects at the oxide-substrate interface and interferes with the direct upward wave. When the thickness of the SiO$_2$ buried oxide layer changed from 900 nm to 1150 nm, the coupling efficiency rose from 22% to 64%. Compared with the long grating coupler, the coupler with a short grating structure was more compact and no focusing lens was needed. It was easy to fabricate with only one additive lithography process. Besides, it had a relatively wide bandwidth (1 dB bandwidth of 40 nm). It was compatible with CMOS manufacturing processes and easy to arrange with the single-mode fiber for on-wafer or on-chip tests.

To increase light coupling efficiency, it is necessary to enhance the grating directionality. One technical solution is to increase the light reflection at the interface in order to reduce the intensity of the downward-propagating wave. The research group of Ghent University proposed and demonstrated a high-efficiency grating coupler by using a bottom metal mirror as a reflector [26]. Benzocyclobutene (BCB), a low-index (n = 1.54 at λ = 1.55 μm) polymer, was chosen as the buffer layer. The optimum simulated BCB buffer thickness was 840 nm. Its thickness was optimized in order to get a constructive interference between the directly upward-radiated wave and the reflected wave at the bottom mirror. The SOI structure, covered with a metal mirror, was bonded onto a host substrate with another BCB layer, and finally, the Si substrate was removed. Compared with a conventional grating coupler, it was an upside-down structure and the grating directionality was almost 100% due to the perfect reflector of the bottom metal mirror. The measured light coupling efficiency to an optical fiber was 69%. The theoretical coupling efficiency was 90% after optimization of etched depth and grating period. However, the bottom metal mirror added more fabrication complexity, which was not compatible with CMOS fabrication processes.

The directionality of the grating diffraction can be improved by changing the structures of its grating as well as increasing the reflection from the substrate. The directionality of the upward diffraction is closely related to the thickness of the waveguide and the top silicon layer of SOI. A simulation had shown that a 370 nm thick waveguide and a 200 nm etch depth could realize enhanced directionality in SOI grating couplers.

Figure 5.11 shows a kind of grating coupler with a polysilicon layer reported by Ghent's group [27]. An additional polycrystalline silicon layer was deposited locally on the SOI substrate and the grating coupler was formed by etching the poly-Si layer as well as the top Si layer of the SOI substrate. The additive poly-Si layer modified the grating structure and enhanced the upward-propagating wave created by the grating. Therefore, it improved the grating directionality. The simulated coupling efficiency and 3 dB bandwidth of this structure were 78% and 85 nm, respectively. The measured coupling efficiency with a single-mode fiber was as high as 55% at 1550 nm [28]. It demonstrated that the coupler with a polysilicon layer enhances the grating directionality and thus raises the coupling efficiency. Furthermore, it is compatible with CMOS fabrication processes.

Conventional grating couplers are slightly inclined with respect to the vertical axis of the wafer surface. This prevents a large second-order back reflection into the SOI waveguide, which reduces the coupling efficiency. Both simulations and experiments have proven that the coupling efficiency of a tilted optical fiber can be 10% higher than that of vertical light coupling.

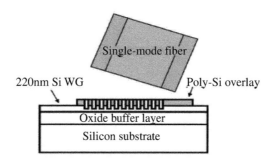

Figure 5.11 The grating coupler with a polysilicon layer. Reprinted with permission from Ref. [27] © The Optical Society.

OEIC chips are mounted in ceramic or plastic housings. Electrical contacts lead outside to contact pins, like in standard microelectronics packages. Optical contacts need fiber throughputs that are preferred for standard packaging in the horizontal or vertical direction. Vertical coupling with high directionality is a specific target mainly taking into account simple standard packaging.

A kind of vertical grating coupler with an extra slit was proposed and demonstrated [29]. Extra deep slits were fabricated by an additive lithography process after the ordinary grating structure was etched. The deep slits could suppress the second-order Bragg reflection from the grating. Theoretically, the couplers based on deep slits have a broad bandwidth and high coupling efficiency of 80% but the additive lithography process needed for the extra deep slit should be carefully designed and fabricated to achieve a high coupling efficiency. The short distance between the slit and the grating structure added complexity to fabrication processes.

A new kind of grating coupler made by relatively simple fabrication processes could obtain vertical light coupling [30]. Focused ion beam (FIB), which was used for a slanted grating structure, is a common tool for electronic device investigation and trimming. Use of gallium ions is standard in FIB. FIB is also useful for nanophotonics as it can be used to create complex 3D structures and accurate modification for nanostructures. However, it can introduce lattice damage and thereby high optical loss. Even the FIB with a low implantation dose causes an additive optical loss of 0.2–2 dB.

The vertical grating coupler garnered a lot of research interest due its high coupling efficiency, broad bandwidth, easy approach for integration, and compatibility with CMOS fabrication processes.

5.4.2 Horizontal Dual Grating–Assisted Coupler

Grating couplers can realize not only vertical light coupling but also horizontal light coupling between an optical fiber and a nanophotonic waveguide. An SOI dual grating–assisted directional coupler was reported by the group of Surrey [34]. The top layer was the SiON waveguide of a large dimension, which matched the optical fiber. The middle layer was the Si_3N_4 waveguide, which was used as a bridge between the SiON waveguide and the bottom SOI waveguide of a small dimension. This waveguide structure was fabricated one layer after another, and two grating structures were used for the light coupling between them. The refractive index of the SiON is 1.478, close to that of the optical fiber, resulting in less than a 0.05 dB insertion loss caused by a refractive index mismatch between the fiber and the waveguide. A fiber was butt-coupled to the thick SiON waveguide, and then the light was coupled into the Si_3N_4 waveguide using the first grating and subsequently coupled into the SOI waveguide via the second grating. The Si_3N_4 waveguide was important for highly efficient light coupling as it bridged the gap between SiON and the Si layer in terms of both refractive index and thickness. A thin SiON cladding layer was deposited on the Si_3N_4 and SOI waveguide to prevent leakage loss. Unibond 4-inch silicon wafers with a 3 µm buried SiO_2 layer and a 230 nm silicon top layer were used for the fabrication of this device. Plasma-enhanced chemical vapor deposition was used for the fabrication of SiON and Si_3N_4 layers. The silicon waveguide was fabricated by ICP etching, and the top three layers were structured by CHF_3 etching. A scanning electron micrograph of this device showed that the total height of SiON and Si_3N_4 layers suffered from overetching and therefore the maximum theoretical efficiency fell from 93% in the original design to 60% for the device as fabricated.

It is worth noting that for fabricating this horizontal dual grating–assisted directional coupler:

- The composition and thickness of the SiON should be carefully designed to match with the optical fiber and to reduce the insertion loss.
- The thickness of the intermediate Si_3N_4 should also be controlled carefully so that SiON and SOI waveguides match each other.
- The parameters of the grating structures, such as etch depth, length, and grating period, should be carefully designed to achieve a small device size and high coupling efficiency.

For fabrication, the growth of a multilayer structure and etching grating structure on the small-dimension waveguide are very difficult. Besides, the horizontal mode mismatch can cause an additive insertion loss.

The dual grating–assisted directional coupler is polarization insensitive when waveguide dimensions are properly designed and elastic stress is carefully controlled.

5.5 Conclusion

On a chip, the waveguide dimensions are small (submicron) whereas in the outer world, fiber waveguides dominate light transmission. The core of the fiber varies from about 10 µm diameter for a monomode fiber to 100 µm and more for a multimode fiber. The simplest butt coupling by horizontal alignment of fiber and nano-waveguide is very ineffective due to size and refractive index mismatch. Direct butt coupling is only used for qualitative experiments, and even then, it requires precise mechanical alignment of the fiber position. High-efficiency coupling needs special coupler structures. All the approaches for high-efficiency light coupling, including taped and inverted tapered spot-size converters, prism couplers and inverted prism couplers, and vertical and horizontal grating couplers, have been investigated widely, and from 2002 to 2008, the basic structures in SOI with good properties were demonstrated.

There are various kinds of silicon light couplers that have their own advantages, and they should be employed as building blocks integrated with other photonic devices according to their fabrication accuracy and complexity, coupling efficiency, and bandwidth. For the tapered spot-size converters, antireflection coatings are required to prevent back reflection from the facets. The taper should operate adiabatically; that is, the local first-order mode of the waveguide should propagate through the taper, while undergoing relatively little mode conversion to higher-order modes or radiation modes [35]. The inverted tapered spot-size converters have relatively small sizes, a high alignment tolerance, broad bandwidths, and high coupling efficiency. However, they are polarization sensitive, which hampers their application. For integration of waveguide devices and SOI active devices, inverted tapered spot-size converters of high coupling efficiency and relatively simple fabrication process are a good choice [36, 37]. A coupling efficiency of more than 90% can be achieved. The measured coupling efficiency of light coupling between an inverted tapered spot-size converter and a rib waveguide was raised to 0.7 dB by NTT [38].

Theoretical simulation shows very high coupling efficiency for grating couplers, but the structures of vertical grating couplers are complex. However, fabrication is possible using standard microelectronics process steps and planar surfaces without additional steps, such as facet polishing or bonding. The performance was improved to nearly perfect coupling within a decade. Manfred Berroth and coworkers at the University of Stuttgart give a good overview [39] about this development. They demonstrated a structure with a record measured efficiency of –0.62 dB at a wavelength of 1531 nm and a 1 dB bandwidth of 40 nm. The performance is enhanced through a backside metal mirror to enhance the directionality and an aperiodic grating that diffracts the field in a Gaussian-like form, improving overlap with the fiber mode.

Acknowledgments

We thank Hao Xu and Yu Zhu for the collection of basic information and figure drawings.

References

1. R. A. Soref and J. P. Lorenzo (1986). All silicon active and passive guided-wave components for λ = 1.3 μm and 1.6 μm, *IEEE J. Quantum Electron.*, **22**, 873–879.
2. Y. A. Vlasov and S. J. McNab (2004). Losses in single-mode silicon-on-insulator strip waveguide and bends, *Opt. Express*, **12**, 1622–1631.
3. L. Vivien, S. Laval, B. Dumont, S. Lardenois, A. Koster and E. Cassan (2002). Polarization-independent single-mode rib waveguides on silicon-on-insulator for telecommunication wavelengths, *Opt. Commun.*, **210**(1), 43–49.
4. S. J. McNab, N. Moll and Y. A. Vlasov (2003). Ultra-low loss photonic integrated circuit with membrane-type photonic crystal waveguides, *Opt. Express*, **11**, 2927–2939.
5. D. Taillaert, P. Bienstman and R. Baets (2004). Compact efficient broadband grating coupler for silicon-on-insulator waveguides, *Opt. Lett.*, **29**, 2749–2751.
6. I. Day, I. Evans, A. Knights, F. Hopper, et al. (2003). Tapered silicon waveguides for low insertion loss highly-efficient high-speed electronic variable optical attenuators, *Optical Fiber Communications Conference*, 249–251.
7. Y. Li, J. Yu and S. Chen (2005). Rearrangeable nonblocking SOI waveguide thermooptic 4×4 switch matrix with low insertion loss and fast response, *IEEE Photonics Technol. Lett.*, **17**, 1641–1643.
8. A. Sure, T. Dillon, J. Murakowski, C. Lin, D. Pustai and D. W. Prather (2003). Fabrication and characterization of three-dimensional silicon tapers. *Opt. Express*, **11**, 3555–3561.

9. T. Brenner, W. Hunziker, M. Smit, M. Bachmann, G. Guekos and H. Melchior (1992). Vertical InP/InGasAsP tapers for low-loss optical fiber-waveguide coupling, *Electron. Lett.*, **28**, 2040–2041.

10. M. Chien, U. Koren, T. L. Koch, B. I. Miller, M. G. Young, M. Chien and G. Raybon (1991). Short cavity distributed Bragg reflector laser with an integrated tapered output waveguide, *IEEE Photonics Technol. Lett.*, **3**, 418–420.

11. T. Brenner and H. Melchior (1993). Integrated optical modeshape adapters in InGaAsP/InP for efficient fiber-to-waveguide coupling, *IEEE Photonics Technol. Lett.*, **5**, 1053–1056.

12. G. Muller, G. Wender, L. Stoll, H. Westermeier and D. Seeberger (1993). Fabrication techniques for vertically tapered InP/InGasAsP spot-size transformers with very low loss, *Proc. Eur. Conf. Integrated Optics*, Neuchatel, Switzerland.

13. B. Jacobs, R. Zengerle, K. Faltin and W. Weiershausen (1995). Verticaly tapered spot size transformers by a simple masking technique, *Electron. Lett.*, **31**, 794–796.

14. L. Pavesi and D. J. Lockwood (2004). *Silicon Photonics* (Springer, Berlin).

15. www.confluentphotonics.com/technology/technical_papers.php#fabrication

16. J. J. Fijol, E. E. Fike, P. B. Keating, D. Gilbody, J. LeBlanc, S. A. Jacobson, W. J. Kessler and M. B. Frish (2003). Fabrication of silicon-on-insulator adiabatic tapers for low loss optical interconnection of photonic devices, *Proc. SPIE*, **4997**, Photonics Packaging and Integration III, https://doi.org/10.1117/12.479371.

17. V. R. Almeida, R. R. Panepucci and M. Lipson (2003). Nanotaper for compact mode conversion, *Opt. Lett.*, **28**, 1302–1304.

18. T. Shoji, T. Tsuchizawa, T. Watanabe, K. Yamada and H. Morita (2002). Low loss mode size converter from 0.3 μm square Si waveguides to singlemode fibres, *Electron. Lett.*, **38**, 1669–1700.

19. S. J. McNab, N. Moll and Y. A. Vlasov (2003). Ultra-low loss photonic integrated circuit with membrane-type photonic crystal waveguides, *Opt. Express*, **11**, 2927–2939.

20. Y. Liu and J. Yu (2007). Low-loss coupler between fiber and waveguide based on silicon-on-insulator slot waveguides, *Appl. Opt.*, **46**, 7858–7861.

21. S. H. Tao, J. Song, Q. Fang, M. Yu, G. Lo and D. Kwong (2008). Improving coupling efficiency of fiber-waveguide coupling with a double-tip coupler, *Opt. Express*, **16**, 20803–20808.

22. Z. Lu and D. W. Prather (2004). Total internal reflection–evanescent coupler for fiber-to-waveguide integration of planar optoelectronic devices, *Opt. Lett.*, **29**, 1784–1750.

23. S. Janz, B. Lamontagne, A. Delage, A. Bogdanov, D. X. Xu and K. P. Xu (2005). Single layer a-Si GRIN waveguide coupler with lithographically defined facets, *IEEE International Conference on Group IV Photonics*, 129–131.

24. D. Taillaert, W. Bogaerts, P. Bienstman, T. Krauss, P. Van. Daele, I. Moerman, S. Verstuyft, K. De Mesel and R. Baets (2002). An out-of-plane grating coupler for efficient butt-coupling between compact planar waveguides and single-mode fibers, *IEEE J. Quantum Electron.*, **38**, 949–995.

25. D. Taillaert, R. Baets, P. Dumon, W. Bogaerts, D. Van Thourhout, B. Luyssaert, V. Wiaux, S. Beckx and J. Wouters (2005). Silicon-on-insulator platform for integrated wavelength-selective components, *Proc. IEEE/LEOS Workshop Fibres and Optical Passive Components*, Italy, 115–120.

26. F. Laere, G. Roelkens, M. Ayre, J. Schrauwen, D. Taillaert and R. Baets (2007). Compact and highly efficient grating couplers between optical fiber and nanophotonic waveguide, *J. Lightwave Technol.*, **12**, 151–156.

27. G. Roelkens, D. Thourhout and R. Baets (2006). High efficiency Silicon-on-Insulator grating coupler based on a poly-silicon overlay, *Opt. Express*, **12**, 11622–11630.

28. G. Roelkens, D. Thourhout and R. Baets (2008). High efficiency diffraction grating coupler for interfacing a single mode optical fiber with a nanophotonic silicon-on-insulator waveguide circuit, *Appl. Phys. Lett.*, **92**, 131101–131103.

29. G. Roelkens, D. Thourhout and R. Baets (2007). High efficiency grating coupler between silicon-on-insulator waveguides and perfectly vertical optical fibers, *Opt. Lett.*, **32**, 1495–1497.

30. J. Schrauwen, F. Laere, D. Thourhout and R. Baets (2007). Focused-ion-beam fabrication of slanted grating couplers in silicon-on-insulator waveguide, *IEEE Photonics Technol. Lett.*, **19**, 816–818.

31. T. Ang, G. Reed, A. Vonsovici, A. Evans, P. Routly and M. Josey (2000). Effects of grating heights on highly efficient unibond SOI waveguide grating couplers, *IEEE Photonics Technol. Lett.*, **12**, 59–61.

32. J. K. Butler, N. H. Sun, G. A. Evans, L. Pang and P. Congdon (1998). Grating-assisted coupling of light between semiconductor and glass waveguides, *IEEE J. Lightwave Technol.*, **16**(6), 1038–1048.

33. R. Orobtchouk, N. Schnell, T. Benyattou, J. Gregoire, S. Lardenois, M. Heitzmann and J. M. Fedeli (2003). New ARROW optical coupler for optical interconnect, *IEEE International Conference on Interconnect Technology*, USA, 233–235.

34. G. Z. Masanovic, V. M. N. Passaro and G. T. Reed (2003). Dual grating-assisted directional coupling between fibres and thin semiconductor waveguides, *IEEE Photonics Technol. Lett.*, **15**, 1395–1397.

35. Y. Fu,, T. Ye, W. Tang and T. Chu (2014). Efficient adiabatic silicon-on-insulator waveguide taper, *Photon. Res.*, **2**, A41–A44.

36. K. K. Lee, D. R. Lim, D. Pan, C. Hoepfner, W. Oh, K. Wada, L. C. Kimerling, K. P. Yap and M. T. Doan (2005). Mode transformer for miniaturized optical circuits, *Opt. Lett.*, **30**, 498–500.

37. M. Galarza, D. Van Thourhout, R. Baets and M. Lopez-Amo (2008). Compact and highly-efficient polarization independent vertical resonant couplers for active-passive monolithic integration, *Opt. Express*, **16**, 8350–8358.

38. T. Tsuchizawa, K. Yamada, T. Watanabe, H. Fukuda, H. Nishi, H. Shinojima and S. Itabashi (2008). SSCs for rib-type silicon photonic wire waveguides, *Proc. IEEE Intern. Conf. Group IV Photonics*, Tokyo, 200–202.

39. W. Sfar Zaoui, A. Kunze, W. Vogel, M. Berroth, J. Butschke, F. Letzkus and J. Burghartz (2014). Bridging the gap between optical fibers and silicon photonic integrated circuits, *Opt. Express*, **22**, 1277–1286.

Chapter 6

Photonic Crystals

6.1 Introduction

Photonic crystals (PCs) are optical materials [1, 2] with periodic changes in the dielectric constant on the wavelength scale, in which forbidden propagation frequency ranges called photonic bandgaps (PBGs) can be created for certain ranges of photon energies. In PCs, we can describe photons in terms of a band structure, as in the case of electrons. Depending on the dimensions of the periodic array, 1D, 2D, and 3D PCs can be realized. A dielectric Bragg mirror, which is well known for many different applications, is an example of 1D PC. In Fig. 6.1, the 1D, 2D, and 3D PC structures created by intersecting two different materials with high and low refractive indexes are presented.

Light of frequencies corresponding to PBG cannot be transmitted in a PC except when defects are introduced. If a defect is included in the PBG structure, a state can be formed in the gap. This state is analogous to a defect or impurity state in a semiconductor that forms a level within the semiconductor bandgap. Defects in PBG structures are engineered by incorporating breaks in the periodicity of the PBG device. A break in periodicity leads to a defect state whose shapes and properties would be dictated by the defect's nature.

Silicon-Based Photonics
Erich Kasper and Jinzhong Yu
Copyright © 2020 Jenny Stanford Publishing Pte. Ltd.
ISBN 978-981-4303-24-8 (Hardcover), 978-981-4303-25-5 (eBook)
www.jennystanford.com

A defect in a PC could, in principle, be designed to be of any size, shape, or form and could be chosen to have any of a wide variety of dielectric constants. Thus, defect states in the gap could be tuned to any frequency and spatial extent of design interest.

Figure 6.1 1D, 2D, and 3D PCs by alternating of refractive index in corresponding dimensions.

A PC being a kind of artificial material, its properties depend on the physical characteristics of the elementary materials that go into its construction. In this case, silicon shows a lot of advantages in realizing PC structures in the following two aspects: firstly, the index contrast between silicon and silica or air is very large, which is vital to a PC because a large enough index modulation is the premise of creating a wide enough PBG. Second, the feature size of a PC is on the order of wavelength, which means we need to fabricate an index-modulated structure comparable to the working wavelength, which is not a problem when using mature silicon-based processes. Therefore, silicon PC possesses theoretical and real possibilities to discover and apply excellent and special photonic effects of PC on subwavelength and nanoscale, which has made it a hot research topic within the last decades, and a lot of novel and high-performance devices and circuits have been created based on silicon PC.

Ideally, a complete PBG exists in a 3D PC. However, due to the difficulty of 3D fabrication to date, most of the experimental research and breakthrough in silicon-based PCs is based on 2D PCs. In the following parts, we are focusing on the simulation, fabrication, and characterization of silicon-based PC slabs owing to the said reason.

6.2　Master Equation

Maxwell's equations have been employed to analyze such periodic structures. In mks units, we have:

$$\nabla \times \bar{D}\rho$$
$$\nabla \times \bar{D} = 0$$
$$\nabla \times \bar{E} = -\frac{\partial \bar{B}}{\partial t} \tag{6.1}$$
$$\nabla \times \bar{H} = \frac{\partial \bar{D}}{\partial t} + \bar{j}$$

Here $\bar{E}$ and $\bar{H}$ are electric and magnetic fields, $\bar{D}$ and $\bar{B}$ are displacement and magnetic induction fields, and it is assumed that free charges and electric current are absent.

For linear, lossless, and isotropic dielectric materials, the following relationships are well known: the electric field and the displacement can be related by Eq. 6.2:

$$\bar{D} = \varepsilon_0 \varepsilon \bar{E}$$
$$\bar{B} = \mu_0 \mu \bar{H} \tag{6.2}$$

Here ε and ε_0 are the dielectric constants of the PC material and free space, respectively and μ and μ_0 are the magnetic permeability of the PC material and free space, respectively. For most of the PC materials of interest, $\mu = 1$ and $p = j = 0$.

By substituting Eq. 6.2 into Eq. 6.1, Maxwell's equations become a group of linear equations. $\bar{E}$ and $\bar{H}$ can be written as harmonic modes, and the time and spatial dependence of $\bar{E}$ and $\bar{H}$ can be separated.

$$\bar{H}(r,t) = \bar{H}(r)e^{-i\omega t}$$
$$\bar{E}(r,t) = \bar{E}(r)e^{-i\omega t} \tag{6.3}$$

The following equation can be obtained [3] by substituting Eq. 6.3 into Eq. 6.1:

$$\Delta \times \left(\frac{1}{\varepsilon(r)} \Delta \times \bar{H}(r) \right) = \left(\frac{\omega}{c} \right)^2 \bar{H}(r) \tag{6.4}$$

Here c is the velocity of light in free space $(c = 1/\varepsilon_0 \mu_0)$. Equation 6.4 is called the master equation of a PC giving out the distributions of magnetic field $\bar{H}(r)$ at the corresponding frequency ω.

6.3 Calculation Methods

6.3.1 PWE Method

For a PC structure with a periodic dielectric material array, the electromagnetic modes have to satisfy not only the master equation but also a periodic condition. Because of the periodicity of a PC structure, the PC material system is unchanged for the translation along the direction of the periodic array, which is called a translational symmetry system and d denotes the lattice vector. Electromagnetic modes for a translational symmetry system [4] can be written in a "Bloch form." In this case, Eq. 6.4 can be expanded with a group of planar waves.

Hence, bandgap maps and the inside optical field can be calculated by using Bloch theory to solve Maxwell's equations with a periodic dielectric distribution.

As an example, we calculate the PBG of a silicon-on-insulator (SOI) PC slab with $t = 0.5a$ and $r = 0.38a$, where a, r, and t represent the lattice constant of PC, the radius of air holes, and the thickness of the top silicon waveguide layer, respectively. The result is illustrated in Fig. 6.2. It is seen that for transverse electric (TE)-like modes, the PBG is from a normalized frequency of 0.3078 to 0.4415 and there

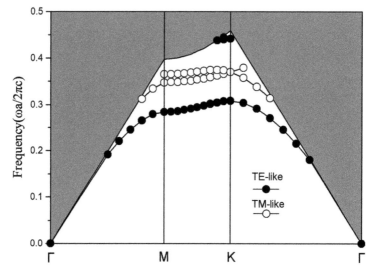

Figure 6.2 Calculation of an SOI PC slab with PWE.

is no PBG for transverse magnetic (TM)-like modes. Therefore, if the center wavelength is 1.3 µm, then a = 480 nm, r = 182.4 nm, and t = 240 nm are proper dimensions of the PC. The calculated bandgap corresponds to the wavelength range of 1087 nm to 1559 nm, which is wide enough for fiber communication applications.

However, what has to be clarified here is that the above planar wave expansion (PWE) method suitable for the perfect PC structure should be modified for PC structures with defects, because a lot of photonic effects are only meaningful in a PC structure with defects. For the simulation of such structures, we need to use the so-called supercell method. And the point is that the supercell but not the original Bragg cell constructs the whole crystal. For example, if we want to calculate a PC line defect waveguide as shown in Fig. 6.3a, the rectangular region that contains the repeatable defect structure is taken as the supercell used in calculation instead of the original square cell. And in the reciprocal lattice, the range of the first Brillouin zone changes from YMX to Y'M'X'. Detailed theoretical methods and considerations to calculate PC modes using the PWE are available in the literature.

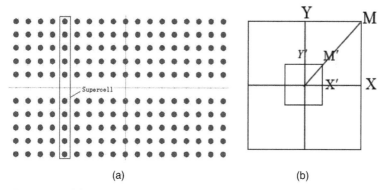

Figure 6.3 (a) The supercell used in the calculation of the PC line defect waveguide. (b) YMX and Y'M'X' correspond to the first Brillouin zone of the original Bragg cell and supercell.

6.3.2 FDTD Method

A finite difference time domain (FDTD) method solves Maxwell's equations numerically [5, 6].

The derivatives in the above equations are approximated by finite differences and the electromagnetic field components are located on a Yee cell. For 2D FDTD, the electric field components at time $n\Delta t$ are located on the sides of the Yee cell while the magnetic field components at times $(n + 1/2)\Delta t$ are located at the center of the Yee cell, as shown in Fig. 6.4 for TE mode.

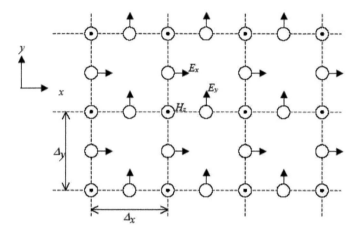

Figure 6.4 Yee cell used in FDTD for TE mode.

Two other problems should be taken into account when using FDTD. One is the algorithm stability. The Courant condition should be satisfied in order to avoid numerical instability. For a given grid size, the time step should be less than a certain value given by the Courant condition. The grid size used for a simulation should be small enough to resolve the smallest feature in the fields and structure during a simulation.

The other problem is the boundary condition. The boundary conditions at the spatial edges of the computational domain must be carefully considered. Many simulations employ an absorbing boundary condition that eliminates any outward propagating energy that impinges on the domain boundaries. One of the most effective is the perfectly matched layer (PML), in which both electric and magnetic conductivities are introduced in such a way that the wave impedance remains constant, absorbing the energy without inducing reflections. In other words, a PML boundary, which consists of several points at the edge of the domain, is designed to act as a highly lossy material, which absorbs all incident energy without producing reflections. This allows field energy that hits the boundary to effectively leave the domain.

6.4 Silicon-Based PC Slab

6.4.1 Important Points about the SOI PC Slab

In a pure 2D PC, the electric field of TM polarization is along the axes air hole or dielectric rods forming PC and the electric field of TE polarization is perpendicular to that of the TM counterpart. However, in a quasi-2D structure, for example, a PC slab as shown in Fig. 6.5, because of alternation of EM boundary conditions and breakage in symmetry, we cannot get pure TE or TM modes but TE-like or TM-like modes with a part of electric fields in other directions. In this case, there may only be a bandgap in one symmetry/polarization and the coupling of TE-like and TM-like modes may lead to extra loss in PC devices.

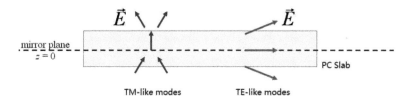

Figure 6.5 Symmetry in a PC slab.

Calculation shows that for a PC slab constructed by dielectric rods in air, a PBG exists only for TM-like polarization. And for a PC slab constructed by air holes in dielectric material, a PBG exists only for TE-like polarization. Considering the coupling loss to air for an air hole PC is smaller than for a rod PC, the former is used more often to design and fabricate devices.

To realize an air hole PC structure in an SOI wafer, normally four structures can be fabricated as shown in Fig. 6.6. Figure 6.6a is the basic structure, which is the easiest to fabricate. However, because of an asymmetric structure (air/Si/SiO$_2$), TE-like mode and TM-like mode are much easier to couple, leading to a larger loss in waveguides and microcavities. If we etch the air holes down to buried SiO$_2$ on the basis of Fig. 6.6a, the coupling loss can be made less and the PBG becomes wider owing to a larger effective index contrast between core and cladding. To get a symmetric structure, we can either remove the buried SiO$_2$ to form an PC air-bridge (Fig. 6.6b) or deposit a layer of SiO$_2$ on top of the original structure. The air-bridge has the best optical performance, but it is complicated in process and easy to collapse. The last structure is also symmetric, but the refractive index contrast between Si and SiO$_2$ is less than that between Si and air and the resulting PBG is narrower. And the quality of coated SiO$_2$ also affects device performance. No matter which kind of PC structure of Fig 6.6 is adopted, the thickness of the core should be less than the single-mode height of the same sandwiched planar waveguide since vertical confinement is achieved by index contrast as discussed above and there is a certain value of core thickness for achieving the widest PBG for all structures.

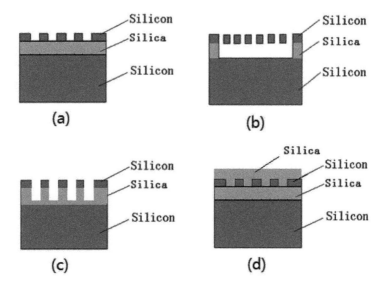

Figure 6.6 Four different PC structures in an SOI slab: (a) asymmetric slab, (b) air-bridge, (c) deep-etched asymmetric slab, and (d) symmetric slab with SiO$_2$ coated.

6.4.2 Fabrication of Silicon-Based PC Slab

For the fabrication [7] of 1D or 2D PCs, normally micro- or nanofabrication techniques are used. The basic process flow is listed in Fig. 6.7 and Table 6.1, taking the fabrication of an SOI-based passive PC structure as an example. Given an SOI wafer, first we need to clean it thoroughly to avoid the influence of contaminations on device performance. Then photoresist is spun-on in order to "record" optical lithography or the e-beam lithography result. Since typical dimensions of near-infrared or mid-infrared PCs are about 500 nm, deep UV lithography or e-beam lithography is often used to get higher precision. The thicker the photoresist, the lower the resolution will be in lithography, but the etch ratio of resist over silicon in the followed etching process should be considered when a resist thickness is selected. Before lithography, the resist needs baking to vaporize the solvent. Then lithography and development are done to define the structure in the resist, followed by another baking called "postbake" to drive off the developing solution and to harden the resist for etching. The purpose of etching is to transfer the structure from resist to wafer, and inductively coupled plasma (ICP) or reactive-ion dry etching method is mostly used for silicon. The last step is to remove the resist from the wafer after etching, using oxygen plasma or acetone.

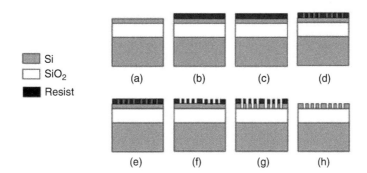

Figure 6.7 Fabrication of SOI passive PC structures [7]: (a) wafer cleaning, (b) resist spin-on, (c) prebake, (d) lithography (e) development, (f) postbake, (g) etching, and (h) resist removing.

132 | *Photonic Crystals*

Table 6.1 Example of a process flow for the fabrication of SOI passive PC structures

No.	Process	Purpose	Recipe/description
(a)	Wafer cleaning	To avoid deterioration of structure owing to contamination	1. Acetone bath (10 min. ultrasonic) 2. Methanol bath (5 min. ultrasonic) 3. Deionized water (DI) rinse 4. RCA clean ($H_2O:NH_4OH:H_2O_2$ = 5:1:1) 5. DI rinse 6. HF dip 7. DI rinse and blow dry (N_2)
(b)	Resist spin-on	To prepare a film to "record" the lithography result	1. Material: 950 K PMMA C_2 2. Parameters: spin-coated at 3000 rpm 3. Result: thickness $\approx$ 150 nm
(c)	Prebake	To vaporize solvent of resist	On a hotplate at 180°C for 10 min.
(d)	Lithography	To define structure in resist	1. Acceleration voltage: 10 kV (e-beam) 2. Aperture: 30 μm 3. Area dose: 70 μA s/cm^2
(e)	Development	To present structure in resist	1. Developed in MIBK:IPA = 1:3 for 15 s 2. Rinsed by IPA for 15 s 3. Blow dry (N_2)
(f)	Postbake	To drive off developing solution and harden resist for etching	On a hotplate at 90°C for 4 min.
(g)	Etching	To transfer the defined structure from resist to substrate	1. Pressure = 1.6 Pa 2. SF6 flow = 60 sccm 3. C4F8 flow = 65 sccm 4. ICP power = 800 W 5. RF power = 50 W
(h)	Resist removing	To obtain the structure ready to be measured	Oxygen plasma under power 300 W or soaking the sample in acetone

Two different samples fabricated with e-beam lithography and ICP etching are shown in Fig. 6.8. Figure 6.8a is an SOI asymmetric PC slab, and Fig. 6.8b is an SOI air-bridge, in which the buried silica is removed with HF solution.

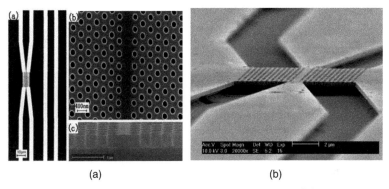

Figure 6.8 (a) SOI PC waveguides in an asymmetric slab and (b) air-bridge.

6.5   SOI PC Devices

6.5.1   SOI PC Waveguides

As a channel of light transmission, an optical waveguide is the basic structure in integrated optics. By making a line defect, we can create an extended mode that we can use to guide light. Traditionally, we achieve waveguiding in dielectric structures, such as optical fibers, by total internal reflection. When the fibers are bent very tightly, however, the angle of incidence becomes too large for total internal reflection to occur and light escapes at the bend. PCs can be designed to confine light even around tight corners because they do not rely on the angle of incidence for confinement.

To illustrate this point, we remove a row of air holes from the PC described earlier. This introduces a defect-guided-mode band inside the gap. The field associated with the guided mode is strongly confined in the vicinity of the defect and decays exponentially in the crystal. An intriguing aspect of PC waveguides is that they provide a unique way to guide optical light, tractably and efficiently, through narrow channels of air. Once light is introduced inside the waveguide, it has nowhere else to go. The only source of loss is reflection from the waveguide input, which suggests that we might use PCs to guide light around tight corners, as in Fig. 6.9. Although the bend radius of curvature is less than the light wavelength, nearly all the light is transmitted through the bend over a wide range of frequencies in the gap. The small fraction of light that is not transmitted is reflected. For specific frequencies, we can achieve 100% transmission.

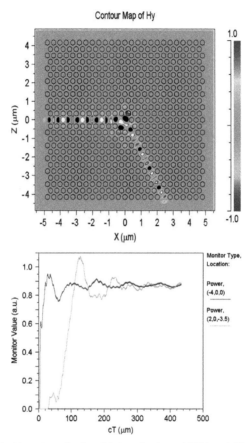

Figure 6.9 An Hy pattern in the vicinity of a sharp 120° bend. The electric field is polarized along the axis of the dielectric rods. The green circles indicate the rod position. Unlike the mechanism of total internal reflection, a PC allows light to be guided in air.

A typical PC line defect waveguide in an SOI slab with transmission characteristics is shown in Fig. 6.10. The air hole radius and periodicity are 135 nm and 410 nm, respectively, fabricated in an SOI slab with a core thickness of 240 nm. It is seen that the stop-band appears at the wavelength of 1590 nm. And for wavelengths longer than this value, PC modes cannot be transmitted, but refractive index-guided modes still exist, so a bit higher transmission is observed as shown in Fig. 6.10.

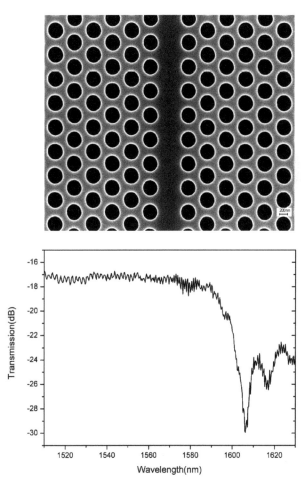

Figure 6.10 PC line defect waveguide in an SOI slab and its transmission characteristics.

6.5.2 SOI PC Microcavities

Electromagnetic resonant cavities [8], which trap light within a finite volume, are an essential component of many important optical devices and effects, from lasers to filters to single-photon sources.

Cavities are characterized by two main quantities: the modal volume V and the quality factor Q. In many applications, high Qs and small Vs are highly desirable for the high finesse required for laser and filter applications and for the high Purcell factor required for controlling the spontaneous emission of atoms placed in resonance with the microcavity mode. The control of spontaneous emission is of interest for increasing the light output and narrowing the linewidth of light-emitting diode structures and for reducing the lasing threshold of semiconductor lasers.

Figure 6.11 illustrates a sample single-defect microcavity in an SOI PC slab together with its resonant spectrum and electric field distribution. Here the air hole radius and periodicity of the triangular

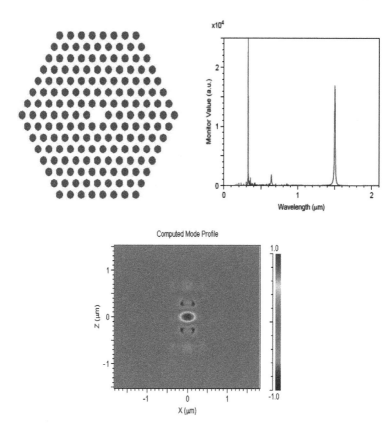

Figure 6.11 Single-dot defect microcavity in an SOI slab and its resonant spectrum and electric field distribution.

lattice are 126 nm and 420 nm, respectively, and the 3D FDTD method with X, Y, and Z step sizes of 30, 10, and 30 nm, respectively, is used for simulation. It is seen that a strong resonant mode at 1550 nm appears inside the PBG and most of the energy is well confined in the dot defect with one air hole removed.

A variety of passive and active optical photonic crystal microcavities have been constructed. It has been shown experimentally and numerically that fine-tuning of the geometry of the holes surrounding the cavity defect may drastically increase the cavity Q/V factor. Many interesting concepts have been used for optimizing the quality factors of the PC microcavities, including symmetry arguments, cancelation of the multipole far-field radiation [9], Bloch wave engineering for increasing the modal reflectivity, and more recently "gently confining" light to avoid radiation. With mature PC microcavities theory and an advanced fabrication process, the experimental Q factor (Table 6.2 [8–12]) of silicon-based PC microcavity got boosted within a decade by a factor of 10,000 and the simulated Q value exceeded 10^7.

Table 6.2 Progress of silicon-based PC microcavity

Year	Structure	$Q(V)$	Institution
1997	1D PC; one hole removed in a ridge waveguide	265 ($V = 0.055$ μm^3)	MIT [8]
2003	2D PC; three holes removed, with nearby holes shifted in an asymmetric slab	45,000 ($V = 0.07$ μm^3)	Kyoto University [9]
2005	2D PC; three holes removed, with nearby holes fine-tuned in an asymmetric slab	100,000 ($V = 0.071$ μm^3)	Kyoto University [10]
2006	2D PC; fine-tuning periodicity in an asymmetric slab	1,000,000 ($V = 1.3$ (λ/n)^3)	Kyoto University [11]
2007	2D PC; heterostructure surrounding microcavity in an asymmetric slab	2,500,000 ($V = 1.4$ (λ/n)^3)	Kyoto University [12]

6.5.3 SOI PC Filters

Because of many excellent photonic effects, a lot of devices, such as source, modulator, switch, and detector. have been created in silicon-based PCs. Here we take a channel drop filter [13] as an example to show how we design, optimize, and fabricate such devices.

The channel drop filter is designed in a PC slab with an air-hole triangular lattice, as shown in Fig. 6.12a. The two PC waveguide buses separated by seven row air holes are both W1 waveguides along the ΓK direction, and the two symmetric microcavities lying between the parallel waveguides are obtained by removing five air holes in the same direction.

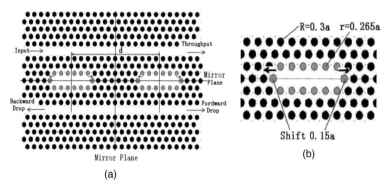

Figure 6.12 (a) Sketch of the optimized PC filter. (b) Structure of the microcavities after optimization.

The radii of the bulk air holes are r = 0.3a, where a is 450 nm, representing the lattice constant. Coupling between the two microcavities results in the formation of two coupled cavity modes, and they are even and odd resonant modes, respectively, with respect to the mirror plane perpendicular to the waveguides. To achieve complete channel drop transfer from one waveguide to the other, the bandwidths and resonant frequencies of the coupled cavity modes must be made equal. The bandwidths of the resonances become equal when the following equation or phase matching condition is satisfied [10]:

$$kd = n\pi \pm \pi/2, \quad (6.5)$$

where k is the wave vector at the operating frequency, d is the distance between the two microcavities, and n is an integer number. As can be seen from Fig. 6.12a, d is chosen to be 13a. Therefore, k should be $0.5\pi/a$ for $n = 6$. From the dispersion curves of the guided modes in the W1 waveguide shown in Fig. 6.13, which was calculated by using the PWE method, it is seen that the corresponding normalized operating frequency for k of $0.5\pi/a$ is about $\omega = 0.2838(a/\lambda)$, as shown by the y axis coordinate of the working point **A** in Fig. 6.13.

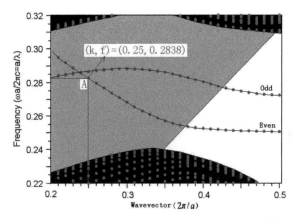

Figure 6.13 The dispersion curves for the guided modes in the W1 waveguide.

To operate the coupled microcavities at the set frequency of $\omega = 0.2838(a/\lambda)$, the cavity structure parameters should be finely tuned. The optimization process was performed by using the 3D FDTD method. The structure of the microcavities after optimization is shown in Fig. 6.12b. It is seen that the radii of the 12 air holes around the defects should be reduced to about $0.265a$ and that the two edge holes of each microcavity should be shifted outward by $0.15a$ from the default positions.

$$1/Q_t = 1/Q_{iso} + 1/Q_c \qquad (6.6)$$

The corresponding wavelength is 1585 nm for $\omega = 0.283(a/\lambda)$ and $a = 450$ nm, and FDTD simulations showed that the Q factor of the isolated microcavity (without coupling with the waveguides) was about 4000. The total Q factor of the filter can be evaluated by using Eq. 6.6, where Q_{iso} is the isolated cavity quality factor, and $1/Q_c$ represents the cavity energy loss due to coupling with the waveguides, which can be calculated by FDTD. Then the total Q factor of the filter that determines the bandwidth was estimated to be about 1260. There are mainly two reasons contributing to energy losses in the microcavities: first, since the working point denoted by A in Fig. 6.13 lies in the light cone of silica, the coupling loss between the resonant modes and the radiation modes in the cladding cannot be avoided. Second, the PC slab is asymmetric, which will result in the coupling of quasi-TE modes and quasi-TM modes in some frequency ranges, and a considerable amount of energy loss arises. If most of the energy losses due to the above mechanisms are suppressed, the total Q factor will be further increased and the maximum can be as high as 15,000, predicted by numerical calculation.

Figure 6.14 shows the electric field $\vec{E}_y$ component in the optimized PC channel drop filter. When the frequency of the input light is off-resonance, the light from port 1 will be transmitted along the W1 waveguide to port 2. However, when the light frequency equals the resonant frequency of the coupled microcavities, the input light will drop to the other W1 waveguide. Since the phase matching equation Eq. 6.20 is satisfied, most of the light will be put out from port 4, as can be seen from Fig. 6.14b. It was calculated that the efficiency of port 4 was about 90%.

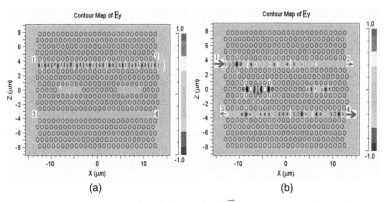

Figure 6.14 3D FDTD-simulated electric field $\vec{E}_y$ component in the PC channel drop filter. (a) Off resonance and (b) on resonance.

The PC filter with the above-optimized parameters was fabricated by electron beam lithography (EBL). First, a 150 nm thick poly(methyl methacrylate) (PMMA) 950 K e-beam resist layer was spun onto a piece of SOI wafer (Soitec Inc.) from a 2% solution in chlorobenzene. The PMMA-layer thickness was chosen in order to avoid electron proximity effect and enable the pattern transfer to the wafer in ICP etching, accounting for the Si/PMMA etching selectivity.

The processing followed the recipe given in Table 6.1. Finally, the wafer was cleaved for measurement and scanning electron microscopy (SEM) characterization.

An SEM image of the central part of the fabricated sample is shown in Fig. 6.15. The access and exit ridge waveguides for measurement are outside the image. It is seen that the air holes have been clearly defined. The air holes were measured to be a little elliptical, which was due to the imperfect adjustment of the EBL system in exposure. The edge roughness of waveguides and air holes is from ±10 to ±20 nm. The above factors lead to a maximum difference of ±10% between the radii of the fabricated and targeted air holes.

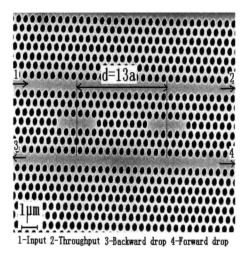

Figure 6.15 An SEM image of the fabricated PC filter on an asymmetric SOI slab.

The transmission spectrum of the sample was measured by using the Newport PM500-C Precision Motion Controller together with the Agilent 8164A Lightwave Measurement System in which a tunable laser (Agilent 81640A), a polarization controller (Agilent 8169A), and a detector unit were integrated. The laser tunable from 1510 to 1640 nm was first modulated to TE-like polarization by Agilent 8169A, which provides polarization synthesis by using a linear polarizer, a quarter-wave plate, and a half-wave plate. The coupling of light with the chip was realized by two identical lensed fibers. The output TE light was selected by a polarizer before it was sent to the built-in detector to ensure measurement of the correct polarization. Though a part of TE light may convert to TM light in the tapered fibers or in the chip, other polarizations were filtered out and only TE light was detected.

The measured spectrum is illustrated in Fig. 6.16. There is a stop-band around the wavelength of 1550 nm, which is due to the PBG effect in the W1 waveguide. A drop peak with an extinction ratio of 6.3 dB appears at about 1598 nm, demonstrating the filter effect of the coupled microcavity system. The drop efficiency of the filter is about 73% ± 5%, estimated by dividing the peak intensity of the "through spectrum" by the peak intensity of the "transmission spectrum" of a reference W1 waveguide at the same wavelength. The total Q factor of the filter is around 1140, through the estimation of the full width at half maximum of the drop peak fitted by a Lorentz function. Therefore, the measured Q factor is very close to the simulation result of 1260. The position of the measured drop peak is red-shifted about 13 nm compared to the simulated result of 1585 nm, owing to the fabrication imperfections as has been analyzed above.

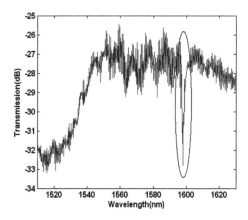

Figure 6.16 The measured spectrum of the filter.

6.6 Conclusions

In conclusion, the basic concepts, calculation methods, fabrication process, waveguides, microcavity, and typical devices of silicon-based PCs are discussed in this section. The invention of PCs makes it possible to manipulate light on a wavelength scale, and the advanced fabrication process and excellent waveguide characteristics of silicon provide an easier way to discover many interesting physical characteristics of PCs, on the basis of which many useful structures and devices can be created. Therefore, silicon-based PCs may be a good choice to achieve high densities in future high-performance photonic integrated circuits.

Acknowledgments

We thank Hejun Yu and Zhiyong Li for the collection of basic information and figure drawings.

References

1. E. Yablonnovitch (1987). Inhibited spontaneous emission in solid-state physics and electronics, *Phys. Rev. Lett.*, **58**(20), 2059–2062.

2. S. John (1987). Strong localization of photons in certain disordered dielectric superlattices, *Phys. Rev. Lett.*, **58**(23), 2486–2489.

3. K. Sakoda (2001). *Optical Properties of Photonic Crystals* (Springer-Verlag, Berlin).

4. J. D. Joannapolous, R. D. Meade and J. N. Winn (1995). *Photonic Crystal-Molding the Flow of Light* (Princeton University Press).

5. K. S. Yee (1966). Numerical solution of initial boundary value problems involving Maxwell's equations in isotropic media, *IEEE Trans. Antennas Propagat.*, **AP-14**, 302.

6. A. Tavlove (1995). *Computational Electrodynamics: The Finite-Difference Time-Domain Method* (Artech House, Norwood).

7. H. Yu, J. Yu, F. Sun, Z. Li and S. Chen (2007). Systematic considerations for the patterning of photonic crystal devices by electron beam lithography, *Opt. Commun.*, **271**(3), 241–247.

8. J. S. Foresi, P. R. Villeneuve, J. Ferrera, E. R. Thoen, G. Steinmeyer, S. Fan, J. D. Joannopoulos, L. C. Kimerling, H. I. Smith and E. P. Ippen (2003). Photonic-bandgap microcavities in optical waveguides, *Nature*, **425**, 944–947.

9. Y. Akahane, T. Asano, B. S. Song and S. Noda (1997). High-Q photonic crystal nanocavity in a two-dimensional photonic crystal, *Nature*, **390**, 143–145.

10. Y. Akahane, T. Asano, B.-S. Song and S. Noda (2005). Fine-tuned high-Q photonic-crystal nanocavity, *Opt. Express*, **13**(4), 1202–1214.

11. T. Asano, B. S. Song and S. Noda (2006). Analysis of the experimental Q factors (-1 million) of photonic crystal nanocavities, *Opt. Express*, **14**(5), 1996–2002.

12. T. Yasushi, H. Hiroyuki, T. Yoshinori, S. Bong-Shik, A. Takashi and S. Noda (2007). High-Q nanocavity with a 2-ns photon lifetime, *Opt. Express*, **15**(25), 17206–17213.

13. H. Yu, J. Yu, Y. Yu and S. Chen (2009). Design and fabrication of a photonic crystal channel drop filter based on an asymmetric Silicon-on-insulator slab, *J. Nanosci. Nanotechnol.*, **9**, 974–977.

Chapter 7

Slow Light in a Silicon-Based Waveguide

7.1 Introduction

Slow light, which means the phenomenon of light propagation in media and structures with a reduced group velocity, has been studied for a long time. It can be traced to the nineteenth century, when the classical theory of dispersion of the electromagnetic waves was first formulated in the works of Lorentz [1]. Slow wave propagation has also been observed and widely used in the microwave range since as early as the 1940s [2, 3].

Going by this history and benefiting from the rapid developments of the technologies required practical implementation of slow light; slow light has become a rapidly growing field with great scientific value and a lot of potential applications. Especially, the recent research on slow light has indicated a variety of potential applications. Examples are given for variable optical delay lines or optical buffers of high-capacity communication networks [4, 5], optical pulse synchronization and reshaping [6], ultrafast all-optical information processing, quantum computing [7], nonlinear optical devices [8], true-time delay in a phased-array antenna [9, 10], optical gyroscope and sensing [11–14], and miniaturization of spectroscopy systems. So far, slow-light propagation has been observed in a

---

Silicon-Based Photonics
Erich Kasper and Jinzhong Yu
Copyright © 2020 Jenny Stanford Publishing Pte. Ltd.
ISBN 978-981-4303-24-8 (Hardcover), 978-981-4303-25-5 (eBook)
www.jennystanford.com

146 | *Slow Light in a Silicon-Based Waveguide*

wide variety of media and structures, including Bose–Einstein condensates, low-pressure metal vapors, solid crystal materials, optical fibers, semiconductor quantum wells and quantum dots, and photonic bandgap structures. Silicon-based optical waveguides are important kinds of slow light schemes because silicon is not only a mature semiconductor device and integrated circuit material but also an excellent photonic material. Silicon-based waveguides, especially SOI-based devices, possess the merits of a small footprint and fabrication compatibility with the complementary metal-oxide-semiconductor (CMOS) technology.

Considerable theoretical and experimental attention to realize slow light helped identify the use of silicon-based waveguides, including silicon microring resonators and silicon-based photonic crystals. In this chapter, we will introduce the foundation of slow light in microring resonators and photonic crystal waveguides (PCWs) and present the important experimental progress.

7.2 Concept

The understanding of slow light is based on the concept of group velocity. Let us recall the distinction between the phase velocity and the group velocity of a light wave.

The phase velocity describes the speed at which the phase of the wave propagates. It can be defined as $v_p = \omega/k = c/n$, where ω is the angular frequency of the light and k is the wave number. And the group velocity gives the velocity with which a pulse of light propagates through a material or structure system. It can be defined as $v_g = \partial\omega/\partial k$. Slow light indicates that the group velocity v_g is much less than the velocity c of light in vacuum. Correspondingly, there is also a concept of fast light (superluminal), when $v_g > c$. The speed of a signal cannot be higher than the phase velocity. A negative group velocity corresponds to the case when the group velocity opposes the phase velocity [15].

Note that the refractive index of the medium $n(k, \omega)$ may be described as a function of the frequency ω of the light as well as a function of the propagation constant k because the dispersion relation connects both quantities. According to the definition of the group velocity and the relationship $k = n\omega/c$, one can get

$$v_g = \frac{c - \omega \dfrac{\partial n(k,\omega)}{\partial k}}{n(k,\omega) + \omega \dfrac{\partial n(k,\omega)}{\partial \omega}} \tag{7.1}$$

From Eq. 7.1, there are two ways to reduce the group velocity v_g as follows: (i) make n larger and (ii) make $\partial n/\partial \omega$ positive. An arbitrary combination of either of the above methods can be used to reduce the group velocity.

The first choice is not favored because the limited range of refractive index variations restricts the possible reductions of the group velocity v_g. Thus, we focus on the ways of making $\partial n/\partial \omega$ or $\partial n/\partial k$ positive and large to realize slow light. Since the functional dependence $n(\omega)$ originates from the material response and the dependence $n(k)$ is best understood from a nonuniform refractive index distribution, we name $\partial n/\partial \omega$ as material dispersion and $\partial n/\partial k$ as waveguide dispersion (or structure dispersion).

The time for a pulse to pass through the optical medium is known as group delay, which can be generally represented as

$$\tau_g = L / v_g, \tag{7.2}$$

where L is the physical length of the propagation medium. When the light field experiences a phase shift of $\phi(\omega)$ as it propagates in a medium, the group delay time can also defined as [16]

$$\tau_g = -\frac{d\varphi(\omega)}{d\omega} \tag{7.3}$$

7.3 Slow Light in Microring Resonator Waveguides

Microring resonators are important components for modern integrated optics that have numerous applications in communication filters [17, 18], optical modulators and switches [19, 20], optical signal processing [21], laser systems [22, 23], and optical sensing [24]. Besides, studies suggest that microring resonators, together with their cascaded structures, are also an excellent slow light element.

For a single microring, there are two types of configuration: all-pass filter (APF) (Fig. 7.1a) and add-drop (Fig. 7.1b) configuration. There are also two main types of cascaded structures: coupled-resonator optical waveguides (CROWs) [25] and side-coupled integrated sequence of spaced optical resonators (SCISSORs) [26, 27]. A CROW is a chain of resonators in which light propagates by virtue of the direct coupling between the adjacent resonators (Fig. 7.1c). In contrast, a SCISSOR consists of a chain of resonators

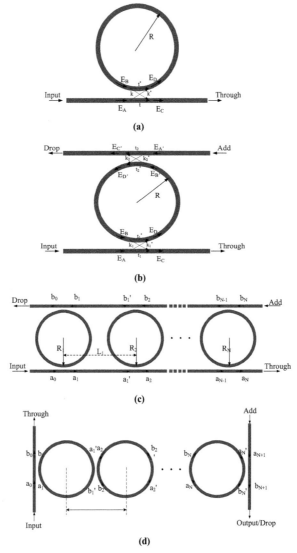

Figure 7.1 The schematic drawing of a single microring and cascaded microrings: (a) all-pass filter (APF) single ring, (b) add-drop single ring, (c) SCISSOR configuration microrings, and (d) CROW configuration microrings.

Slow Light in Microring Resonator Waveguides | 149

not directly coupled to each other but coupled through at least one side-coupled waveguide (Fig. 7.1d). Both CROWs and SCISSORs have the potential to significantly slow down the propagation of light. This part deals with slow light in a single-microring resonator as well as the two types of cascaded structured microring resonator waveguides.

7.3.1 Single-Microring Resonator

Figure 7.1 shows the two basic types of single-ring configuration, APF and add-drop single ring. Using transfer matrixes to analyze the characteristic of the add-drop single ring, we can get the properties of the APF single ring by setting the coupling efficient of the add-drop ring to the drop waveguide to zero.

In the schematic of a add-drop single ring, $E_A(E'_A), E_B(E'_B), E_C(E'_C), E_D(E'_D)$ represent the local optical fields in the waveguide and the ring of the coupling region, and the field relations can be expressed by a transfer matrix. The coupling coefficient κ and transmission coefficient t satisfy the relation $t_i = \sqrt{1 - \kappa_i^2}$ (i = 1, 2). The index i refers to the input line (i = 1) and the drop line (i = 2). We attributed all sources of loss (material absorption, radiation, and scattering) inside the ring to the field transmission γ per round in the ring.

The single-pass phase shift is $\varphi(\lambda) = n_{\text{eff}}(\lambda) \cdot \dfrac{2\pi}{\lambda} \cdot 2\pi R$.

The typical transmission spectra for the trough port (red line) and the drop port (blue line) are plotted in Fig. 7.2.

One can see from the spectra in Fig. 7.2 that when the ring is on-resonance, light is directed to the drop port, and when the ring is off-resonance, light goes to the trough port. When the minimal transmission for the trough port at the resonant wavelength λ_0 equals 0, the ring reaches a critical state that is defined as critical coupling. It means that the coupling between the input waveguide and the ring is equal to the loss per round in the ring times the coupling between the output (drop) waveguide and the ring. This state requires that

$$t_1 = t_2 \cdot e^{-\gamma} \tag{7.4}$$

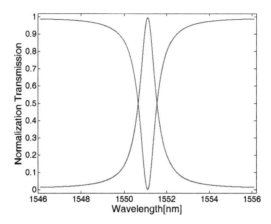

Figure 7.2 Typical transmission spectra of the trough port (red line) and the drop port (blue line) for the microring resonator in the add-drop configuration.

Similarly, $t_1 > t_2 \cdot e^{-\gamma}$ represents the undercoupling state and $t_1 < t_2 \cdot e^{-\gamma}$ represents the overcoupling state, which means the coupling t_1 between the input waveguide and the ring is more or less than the loss $\exp(-\gamma)$ per round in the ring times the coupling t_2 between the output waveguide and the ring, respectively. The coupling state is closely related to the slow light condition, which will be discussed later.

From the transfer matrix, we can also get the effective phase shift at both the trough and the drop port and the group delay time according to the definition Eq. 7.3.

We can get the transfer characteristic, phase shift characteristic, as well as delay characteristic of an APF single ring when the coupling efficiency of the add-drop ring to the drop waveguide is zero.

Let us now discuss the relationship between the coupling and delay characteristics of a single ring. The delay time is a periodic function of wavelength for a fixed coupling states (over- or undercoupling state), which exhibits sharp peaks at $\phi = 2N\pi$. The delay time at $\cos\phi = 1$ is

$$\tau_d \mid_{\lambda_0} = -\frac{(1-t^2)\cdot e^{-\gamma}}{(e^{-\gamma}-t)(1-t\cdot e^{-\gamma})} \cdot \frac{n_g \cdot 2\pi R}{c}. \tag{7.5}$$

At the critical coupling point of the APF single ring, that is, $e^{-\gamma} = t$, the denominator of Eq. 7.5 equals zero. As a result Eq. 7.5 displays a very violent behavior in this region associated with the fact that the effective phase shift changes sign near resonance as coupling crosses the critical point. For $e^{-\gamma} > t$, the ring is overcoupled and the group delay time τ_d has a positive value at resonance, indicating that the light pulse is trapped and spends a relatively long time circulating in the ring and so the ring exhibits a slow light mode. On decreasing $e^{-\gamma}$ while keeping t fixed, τ_d becomes a large positive value as $e^{-\gamma}$ approaches its critical value, then flips to a large negative value as $e^{-\gamma}$ enters the $e^{-\gamma} < t$ region and finally decreases in magnitude (remaining negative) as $e^{-\gamma}$ is further decreased. Figure 7.3 shows the group delay time as a function of wavelength for an APF single ring in different coupling states, in which $t = 0.97$ is kept fixed.

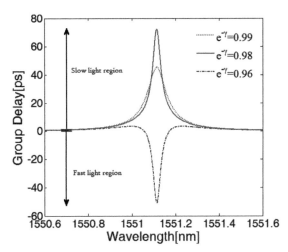

Figure 7.3 Group delay time as a function of wavelength for an APF single ring with a fixed $t = 0.97$.

7.3.2 SCISSOR Configuration Microring Resonators

On the basis of the transfer matrix analysis of a single ring, we can also get the dispersion relation of a SCISSOR. The local optical fields of the SCISSOR are labeled in Fig. 7.1c.

To find the modes and dispersion relation of a SCISSOR configuration, applying the Bloch boundary conditions, we get

$$\begin{bmatrix} a_1' \\ b_1' \end{bmatrix} = \exp(jK\Lambda) \begin{bmatrix} a_0 \\ b_0 \end{bmatrix}, \quad (7.6)$$

where K is the Bloch wave number and $\Lambda = L_1$ is the Bragg period. This relation strictly applies only for an infinite periodic array, but we will see that the photonic band structure exists even for a finite array.

A typical dispersion relation of a SCISSOR structured microring is depicted in Fig. 7.4, from which, one can see that the dispersion relation consists of passbands and bandgaps. Furthermore, the dispersion relation exhibits two types of bandgaps [28]: one is direct gap, which means that there is no discontinuity in the Bloch wave vector between the lower and upper band edge frequencies, and the other is indirect gap, where the wave vectors between the upper and lower band edge frequencies differ by π/Λ. The two types of bandgaps stem from the Bragg reflection and the resonant reflection in SCISSOR structures; thus they are called the "resonant gap" and the "Bragg gap," respectively. If the frequency of the light is close to the Bragg frequency, the additional phase shift induced by successive unit cells of the system is an integer multiple of 2π, and a weak coupling can be enhanced via a Bragg-type process of constructive interference of reflections. Therefore, a Bragg gap arises in the dispersion relation.

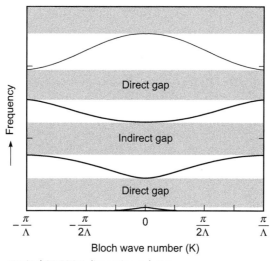

Figure 7.4 Typical SCISSOR dispersion relation.

For a frequency close to the resonance frequency of the resonator, the effective coupling between the two channels can become quite

Slot Light in Microring Resonator Waveguides | 153

large and a gap also arises. This gap is the called the "resonator gap." Near the band edge frequencies, the dispersion relation is flat and the group velocity approaches zero and this is a slow light regime.

7.3.3 CROW Configuration Microring Resonators

There are three methods to analyze the CROW configuration microring: tight-binding formalism [25], transfer matrixes [29], and temporal coupled-mode theory [30]. The three methods of analysis produce an identical form of the dispersion relation in the limit of weak coupling [31]. Here we chose to revisit the matrix analysis above.

In the tight-binding method, we approximate the electric field of an eigenmode $\vec{E}_k$ of the CROW as a Bloch wave superposition of the individual resonator modes.

Substituting the Bloch waves into Maxwell's equations and taking the assumption of symmetric nearest-neighbor coupling, after some algebra, we find that the dispersion relation of the CROW is:

$$\omega_K = \omega_0 [1 - \frac{\Delta\alpha}{2} + \kappa\cos(K\Lambda)] \tag{7.7}$$

Here ω_K is the frequency of the eigenmode in the CROW chain, ω_0 is the resonance frequency of an individual resonator, the coupling parameter κ represents the overlap of the modes of two neighboring resonators, and $\Delta\alpha$ gives the fractional self-frequency shift centered at ω_0.

A dispersion diagram of a CROW-structured microring is shown in Fig. 7.5a [32]. From the dispersion curve, one can see that the propagation waves in CROW are arranged in narrow bands centered at the resonance frequencies, while at the nonresonance frequencies, the optical wave decays in the CROW and forms the bandgaps. The group velocity in a CROW, which corresponds to the slope of the dispersion curve, is given by

$$v_g = \frac{d\omega_K}{dK} = -\omega_0 \Lambda\kappa\sin(K\Lambda). \tag{7.8}$$

From this formula of v_g and the dispersion curve, it can be seen that the group velocity is maximum at the center of the passband, where $\sin(K\Lambda) = 1$, and zero at the band edges, where $\sin(K\Lambda) = 0$. Besides, we can see that the group velocity depends on the coupling efficient κ between the microrings and on the periodicity of the CROW.

To achieve slow-light propagation, the coupling between the microrings must be weak to attain a relatively flat propagation band in the dispersion relation. A small value of group velocity can be acquired for a weakly coupled CROW or compact CROW. Figure 7.5b shows the normalized group velocity as a function of the normalized frequency, which gives a visualization of the characteristics.

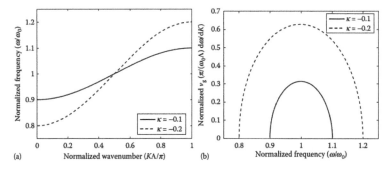

Figure 7.5 (a) Dispersion diagram and (b) the normalized group velocity of a CROW for various values of κ and $\Delta\alpha = 0$. Copyright (2008) From Ref. [32]. Reproduced by permission of Taylor and Francis Group, LLC, a division of Informa plc.

7.3.4 Experimental Progress

With the improvement of microelectronic fabrication and high-index-contrast silicon waveguide technology, ultracompact and large-scale cascaded microrings for slow light have been demonstrated. Besides, to get tunable group delays, some tuning mechanisms, like the thermo-optic effect, were introduced to change the refractive index of the waveguides.

On the basis of the SOI photonic wire waveguides, a research group of IBM realized 56-microring resonators cascaded in SCISSOR configuration and 100 microring resonators cascaded in CROW configuration to achieve a large group delay [33]. The SEM pictures of the optical delay lines are shown in Fig. 7.6. An on-resonance group delay of 510 ps is observed in the time-domain measurements of a 1 Gbps pseudorandom bit stream of optical pulses in a non-return-to-zero format, transmitted through the 56-microring SCISSOR, and the value is 220 ps for the 100-microring CROW.

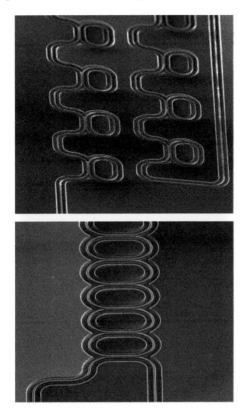

Figure 7.6 Scanning electron micrographs of resonantly enhanced optical delay lines based on photonic-wire waveguides. Reprinted by permission from Springer Nature Customer Service Centre GmbH: Springer Nature, *Nature Photonics*, Ref. [33], copyright (2006).

A balanced SCISSOR delay structure, which allows for an increased bandwidth and continuous tunability of group delay, was demonstrated [34]. In the balanced SCISSOR, the resonant frequencies of the rings are shifted by a small amount $\Delta\omega$ from a center resonant frequency ω_r. Half of the rings are blue-shifted, and the other half are red-shifted compared to the center frequency. This arrangement cancels the third-order group delay dispersion. Even though the resonances of the rings are changed about the central resonant frequency ω_r, the operating frequency of the overall structure remains the same while the group delay significantly changes. The spectral width of the structure, therefore, broadens, increasing its bandwidth. Using standard CMOS fabrication processes, the device consisted of eight SOI-based microrings with Cr heaters. The measurement of the device shows a continuously tunable delay of up to 72 ps operating on 100 ps pulses without introducing distortion, which can be used as on-chip buffers in optical interconnects and signal processing applications.

7.4 Slow Light in Photonic Crystals

PCWs are known for providing large group index dispersion together with very low values of group velocity. A "group velocity refractive index," usually called the "group index" is defined as the ratio of light velocity c to group velocity v_g. This value should not be confused with the refractive index n, which is always defined with respect to the phase velocity.

Slow-light propagation in PCWs is a hot research topic, and group velocities below $c/300$ were early demonstrated [35]. Like slow light in microring resonators, slow light in PCWs is controlled by dispersion from waveguide structures. PCWs offer more bandwidth than any other schemes in a slow light regime. However, loss issues may be one of the challenges for PCW slow light applications.

7.4.1 Generation of Slow Light in a Photonic Crystal Waveguide

A photonic crystal is primarily a grating, and most of the slow light effects can be explained from a 1D grating perspective. In fact, the coupled resonator structures discussed in the preceding section fall into the same category. Photonic crystals are conveniently described by their band structure, which is described as an energy wave vector $(\omega - k)$ diagram of the allowed states or modes of the crystal.

The simplest dispersion curve is shown in Fig. 7.7a, which describes a wave propagating in a dispersion-free medium, namely a straight line with a constant slope. In fact, the constant slope denotes the phase velocity, which is defined as $v_p = \omega/k$. For free space, we have $v_p = c = \omega/k$; for a material of refractive index n, $v_p = c/n = \omega/k$. In a periodically structured medium such as a 1D grating, the dispersion curve is no longer a straight line but now features a discontinuity; see Fig. 7.7b. The discontinuity breaks the dispersion curve into multiple bands consisting of some passbands spaced with stop-bands, namely the PBG features. Only light at the frequency in the passbands can be guided in the structure, while light of the frequency in the stop-band is reflected. In proximity of this stop-band, the dispersion curve is no longer a straight line and one needs to distinguish between the phase velocity and the group velocity $v_g = \partial\omega/\partial k$, the latter denoting the local slope of the curve. From Fig. 7.7b, at the edge of the stop-band, the slope of the dispersion curve is flat, indicating that the group velocity is zero. On the other hand, near the stop-band, the local slope of the curve, namely the group velocity, is larger than zero but much smaller than the phase velocity.

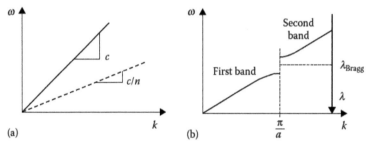

Figure 7.7 Dispersion diagram for (a) an optical wave propagating in free space (solid line) and a medium with a constant refractive index n (dashed line) and (b) an optical wave propagating in a periodic medium. Copyright (2008) From Ref. [32]. Reproduced by permission of Taylor and Francis Group, LLC, a division of Informa plc.

What's the nature of slow light in PCWs? There are two possible mechanisms, backscattering and omnidirectional reflection, to illustrate slow-light propagation in PCWs [36], invoking the familiar ray picture commonly used to describe light propagation in a dielectric waveguide. For backscattering, light is coherently backscattered at each unit cell of the photonic crystal, so the photonic crystal acts as a 1D grating (indicated by the vertical lines on the left in Fig. 7.8). At the Brillouin zone boundary for $k = 0.5 \times 2\pi/a$, the forward propagating and the backscattered light agree in phase and amplitude and a standing wave results, which can also be understood as a slow mode with zero group velocity. If the incident light frequency moves away from the Brillouin zone boundary, it will fall into the slow light regime and the forward and backward traveling components begin to move out of phase but still interact, resulting in a slowly moving interference pattern—the slow mode.

Further from the Brillouin zone boundary, the forward and backward traveling waves are too out of phase to experience much interaction and the mode behaves like a regular waveguide mode that is dominated by total internal reflection. In Fig. 7.8, the arrows pointing right and left represent the forward and backward traveling components, respectively. The right-pointed arrows are longer, as if the mode was taking three steps forward and two steps back—a slow forward movement. Since the existence of the photonic bandgap, light of proper energy that propagates at any angle is reflected. Even light propagating at or near normal incidence may, therefore, form a mode, as indicated by the steep zig-zag on the right in Fig. 7.8. In band structure terms, this corresponds to propagation along the waveguide at or near the Γ point, that is, $k \approx 0$. It is obvious that such

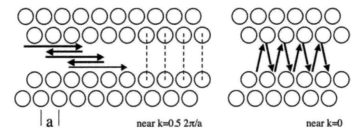

Figure 7.8 Illustration of the two possible mechanisms for achieving slow light in photonic crystal waveguides, namely coherent backscattering (left) and omnidirectional reflection (right). Republished with permission of IOP Publishing, Ltd, from Ref. [36], copyright (2007); permission conveyed through Copyright Clearance Center, Inc.

modes have very small forward components, that is, they travel as slow modes along the waveguide or form a standing wave for $k = 0$.

7.4.2 Experimental Verification of Slow Light in Photonic Crystals

There has been a lot of research work about slow light in PCWs in the last decades, and the covered topics include theoretical exposition [37], experimental observation [38, 39], slow light–enhanced nonlinear effects [40–43], dispersion engineering [44–46], and coupled-nanocavity-enhanced photonic crystals waveguides [47, 48]. Here we review some of the representative works obtained in the early pioneering phase.

Early in 2005, Vlasov et al. had experimentally demonstrated an over 300-fold reduction in the group velocity on a silicon chip via an ultracompact photonic integrated circuit using low-loss silicon PCWs [35]. Figure 7.9 shows the SEM images of the silicon PCWs device they used. To measure the group velocity, an integrated unbalanced Mach–Zehnder interferometer (MZI) structure was designed (Fig. 7.9a). The unbalanced MZI is formed by introducing a small but noticeable difference in the hole radii of the two PCW arms—the signal arm and the reference arm.

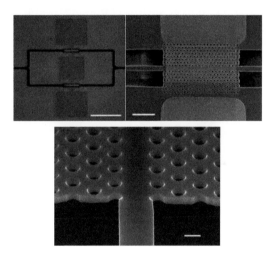

Figure 7.9 SEM images of an active unbalanced Mach–Zehnder interferometer using photonic crystal waveguides. Reprinted by permission from Springer Nature Customer Service Centre GmbH: Springer Nature, *Nature*, Ref. [35], copyright (2005).

Due to the interference between the signal light and the reference light, the frequency becomes nearly independent from the wave number near the bandgap. Very large group indices of up to 500 were found from the transmission spectrum. In addition, optimized ohmic lateral electrical contacts for locally heating the signal arm were fabricated to dynamically tune the dispersion characteristics, thus obtaining active control of slow light. Figure 7.9b shows the definition of the metallic contacts on top of the PCW. Measure of the active MZI device indicates that tunable group indices can be achieved by applying electric power to the signal arm.

Slow light–enhanced nonlinear optical effects have been noticed and utilized. An outstanding representative work is green light emission in silicon through slow-light-enhanced third-harmonic generation (THG) in PCWs [49].

A 2D silicon PCW (see Fig. 7.10) is engineered to display both low group velocity and low dispersion, and the latter feature promises benefits of slow light for nonlinear applications. The measured group velocity of the fundamental mode varies almost linearly by a factor of 4 from $c/10$ to $c/40$ in the spectral window from 1550 to 1559 nm. When a near-infrared 4.5-ps pulse train was launched into the PCW, green light emitting from the surface of the chip was observed by the naked eye (Fig. 7.10). It is illustrated that the green light emission

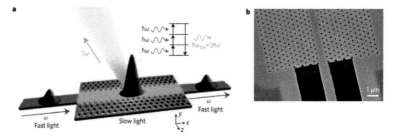

Figure 7.10 Green light emission through third-harmonic generation (THG) in a slow-light photonic crystal waveguide. (a) Schematics of slow light–enhanced THG. The fundamental pulse at a frequency ω is spatially compressed in the slow-light photonic crystal waveguide, increasing the electric field intensity, while the third-harmonic signal, at a frequency $\omega_{TH} = 3\omega$, is extracted out-of-plane by the photonic crystal with a specific angle off the vertical direction. (b) Scanning electron microscopy (SEM) image of the tapered ridge waveguide connected to the photonic crystal waveguide etched in a thin silicon membrane. Scale bar, 1 mm. Reprinted by permission from Springer Nature Customer Service Centre GmbH: Springer Nature, *Nature Photonics*, Ref. [49], copyright (2009).

for only 10 W peak pump powers is due to both the tight light confinement within the PCW and the energy density enhancement provided by the slow light mode. This THG observation further highlights that slow light is favorably utilized to implement the desired functionalities.

7.5 Conclusion

Microring resonators, especially cascaded microring resonators, and photonic crystals are two promising approaches for generating slow light. Progress on both types of structures runs almost in parallel, and the two are competing to become the preferred scheme. Both types of schemes possess potential as well as limits in future application, and each has its own most suitable area of exploitation.

Slow light devices should be evaluated by considering a variety of figures of merit, such as tunability, losses, preservation of the signal quality, reliability, and technology requirements, in addition to the possibility of dynamically controlling the slowdown factor on chip [50].

In recent years, using nonlinear optical effects in hybrid-on silicon emerged as an interesting topic for on-chip all-optical data treatment. Because of their low refractive indexes, the new nonlinear optical materials developed in soft matter science (e.g., polymers and liquids) are delicate for using. However, such materials for using can be incorporated in slot PCWs and hence can benefit from both slow light field enhancement effect and slot-induced ultrasmall effective areas. Reported results [51] provide experimental evidence for accurate control of the dispersion properties of fillable periodical slotted structures [52] in silicon photonics. PCWs appear as good candidates due to their confinement and versatile dispersion properties, including slow light and possible control of group velocity dispersion. Slow light structures [53] show enhancement of the nonlinearity that depends on the group index. Third-order nonlinear effects, such as Kerr self-phase modulation scale, increase with the square of the group index.

Acknowledgments

We thank Yingtao Hu and Yuntao Li for the collection of basic information and figure drawings.

References

1. H. Lorentz (1880). Relation between propagation of light and density of matter, *Wied. Ann. Phys. Chem.*, **9**(4), 641–665.

2. J. R. Pierce (1950). Traveling-wave tubes, *Bell Syst. Tech. J.*, **29**(4), 608–671.

3. L. Field (1949). Some slow-wave structures for traveling-wave tubes, *Proc. IRE*, **37**(1), 34–40.

4. R. W. Boyd (2005). Applications of slow-light in telecommunications and optical switching, Presented at Photonics West.

5. E. Parra and J. Lowell (2007). Toward applications of slow light technology, *Optics and Photonics News*, **18**(11), 40–45.

6. M. Fisher and S. Chuang (2005). Variable group delay and pulse reshaping of high bandwidth optical signals, *IEEE J. Quantum Electron.*, **41**(6), 885–894.

7. M. Fleischhauer and M. Lukin (2002). Quantum memory for photons: dark-state polaritons, *Phys. Rev. A*, **65**(2), 22314.

8. S. Harris and L. Hau (1999). Nonlinear optics at low light levels, *Phys. Rev. Lett.*, **82**(23), 4611–4614.

9. W. Jemison, T. Yost and P. Herczfeld (1996). Acoustooptically controlled true time delays: experimental results, *IEEE Microwave Guided Wave Lett.*, **6**(8).

10. Z. Liu, et al. (2006). X-band continuously variable true-time delay lines using air-guiding photonic bandgap fibers and a broadband light source, *Opt. Lett.*, **31**(18), 2789–2794.

11. F. Zimmer and M. Fleischhauer (2004). Sagnac interferometry based on ultraslow polaritons in cold atomic vapors, *Phys. Rev. Lett.*, **92**(25), 253204.

12. Z. Shi, et al. (2007). Slow-light Fourier transform interferometer, *Phys. Rev. Lett.*, **99**(24), 240804.

13. G. Purves, C. Adams and I. Hughes (2006). Sagnac interferometry in a slow-light medium, *Phys. Rev. A*, **74**(2), 23805.

14. G. Jundt, et al. (2003). Non-linear Sagnac interferometry for pump-probe dispersion spectroscopy, *Eur. Phys. J. D*, **27**(3), 273–276.

15. C. Garrett and D. McCumber (1970). Propagation of a Gaussian light pulse through an anomalous dispersion medium, *Phys. Rev. A*, **1**(2), 305–313.

16. G. Lenz, B. J. Eggleton, C. K. Madsen and R. E. Slusher (2001). Optical delay lines based on optical filters. *IEEE J. Quantum Electron.*, **37**, 525–532.

17. C. Madsen and J. Zhao (1999). *Optical Filter Design and Analysis: A Signal Processing Approach* (John Wiley & Sons, Inc. New York, NY, USA).

18. C. Madsen (2000). General IIR optical filter design for WDM applications using all-pass filters, *J. Lightwave Technol.*, **18**(6), 860.

19. Q. Xu, et al. (2007). 12.5 Gbit/s carrier-injection-based silicon microring silicon modulators, *Opt. Express*, **15**(2), 430–436.

20. S. Emelett and R. Soref (2005). Design and simulation of silicon microring optical routing switches, *J. Lightwave Technol.*, **23**(4), 1800.

21. V. Van, et al. (2002). Optical signal processing using nonlinear semiconductor microring resonators, *IEEE J. Sel. Top. Quantum Electron.*, **8**(3), 705–713.

22. B. Liu, A. Shakouri and J. Bowers (2004). Passive microring-resonator-coupled lasers, *Appl. Phys. Lett.*, **79**, 3564.

23. Z. Bian, B. Liu and A. Shakouri (2003). InP-based passive ring-resonator-coupled lasers, *IEEE J. Quantum Electron.*, **39**(7), 859–865.

24. A. Yalcin, et al. (2006). Optical sensing of biomolecules using microring resonators. *IEEE J. Sel. Top. Quantum Electron.*, **12**(1), 148–155.

25. A. Yariv, et al. (1999). Coupled-resonator optical waveguide: a proposal and analysis, *Opt. Lett.*, **24**(11), 711–713.

26. J. Heebner and R. Boyd (2002). Slow' and fast 'light in resonator-coupled waveguides, *J. Mod. Opt.*, **49**(14/15), 2629–2636.

27. J. Heebner, R. Boyd and Q. Park (2002). SCISSOR solitons and other novel propagation effects in microresonator-modified waveguides, *J. Opt. Soc. Am. B*, **19**, 722–734.

28. J. Heebner (2003). Nonlinear optical whispering gallery microresonators for photonics, PhD thesis, University of Rochester.

29. J. Poon et al. (2004). Matrix analysis of microring coupled-resonator optical waveguides, *Opt. Express*, **12**(1), 90–103.

30. J. Poon and A. Yariv (2007). Active coupled-resonator optical waveguides. I. Gain enhancement and noise, *J. Opt. Soc. Am. B*, **24**(9), 2378–2388.

31. J. Poon, et al. (2004). Designing coupled-resonator optical waveguide delay lines, *J. Opt. Soc. Am. B*, **21**(9), 1665–1673.

32. J. Khurgin (2008). *Slow Light: Science and Applications* (CRC Press).

33. F. Xia, L. Sekaric and Y. Vlasov (2007). Ultracompact optical buffers on a silicon chip, *Nat. Photonics*, **1**(1), 65–74.

34. J. Cardenas, et al. (2009). Large bandwidth continuously tunable delay using silicon microring resonators, in *Group IV Photonics 2009*.

35. Y. Vlasov, et al. (2005). Active control of slow light on a chip with photonic crystal waveguides, *Nature*, **438**(7064), 65–69.

36. T. Krauss (2007). Slow light in photonic crystal waveguides, *J. Phys. D: Appl. Phys.*, **40**, 2666–2670.

37. S. Zhu, et al. (2000). Time delay of light propagation through defect modes of one-dimensional photonic band-gap structures, *Opt. Commun.*, **174**(1–4), 139–144.

38. H. Gersen, et al. (2005). Real-space observation of ultraslow light in photonic crystal waveguides, *Phys. Rev. Lett.*, **94**(7), 73903.

39. T. Kawasaki, D. Mori and T. Baba (2007). Experimental observation of slow light in photonic crystal coupled waveguides, *Opt. Express*, **15**(16), 10274–10284.

40. M. Soljačić, et al. (2002). Photonic-crystal slow-light enhancement of nonlinear phase sensitivity, *J. Opt. Soc. Am. B*, **19**(9), 2052–2059.

41. B. Corcoran, et al. (2009). Optical performance monitoring via slow light enhanced third harmonic generation in silicon photonic crystal waveguides, *Conference on Optical Fiber Communication*, San Diego, CA, USA.

42. Y. Hamachi, S. Kubo and T. Baba (2009). Slow light with low dispersion and nonlinear enhancement in a lattice-shifted photonic crystal waveguide, *Opt. Lett.*, **34**(7), 1072–1074.

43. K. Inoue, et al. (2009). Enhanced third-order nonlinear effects in slow-light photonic-crystal slab waveguides of line-defect, *Opt. Express*, **17**(9), 7206–7216.

44. S. Ha, et al. (2008). Dispersionless tunneling of slow light in antisymmetric photonic crystal couplers, *Opt. Express*, **16**(2), 1104–1114.

45. T. Baba, et al. (2009). Dispersion-controlled slow light in photonic crystal waveguides, *Proc. Jpn. Acad. Ser. B: Phys. Biol. Sci.*, **85**(10), 443–453.

46. M. Ebnali-Heidari, et al. (2009). Dispersion engineering of slow light photonic crystal waveguides using microfluidic infiltration, *Opt. Express*, **17**(3), 1628–1635.

47. M. Notomi, E. Kuramochi and T. Tanabe (2008). Large-scale arrays of ultrahigh-Q coupled nanocavities, *Nat. Photonics*, **2**(12), 741–747.

48. M. Svaluto Moreolo, V. Morra and G. Cincotti (2008). Design of photonic crystal delay lines based on enhanced coupled-cavity waveguides, *J. Opt. A: Pure Appl. Opt.*, **10**, 064002.

49. B. Corcoran, et al. (2009). Green light emission in silicon through slow-light enhanced third-harmonic generation in photonic-crystal waveguides, *Nat. Photonics*, **3**(4), 206–210.

50. A. Melloni, F. Morichetti and T. Krauss (2009). Slow light in coupled ring resonators and PhC: a comparison, *Advances in Optical Sciences Congress*, OSA Technical Digest (CD) (Optical Society of America, 2009), paper STuC3.

51. S. Serna, P. Colman, W. Zhang, X. Le Roux, C. Caer, L. Vivien and E. Cassan (2016). Experimental GVD engineering in slow light slot photonic waveguides, *Sci. Rep.*, **6**, 26956.

52. J. Leuthold, C. Koos and W. Freude (2010). Nonlinear silicon photonics, *Nat. Photonics*, **4**, 535–544.

53. L. Vivien and L. Pavesi (2013). *Handbook of Silicon Photonics* (CRC Press).

Chapter 8

Light Emitters

Silicon is a classic example of an indirect semiconductor (see Chapter 2). The indirect bandgap E_{gind} is created by a transition between band states (valence band edge) of zero wave number ($k = 0$), conventionally called the Γ point, and the lowest lying conduction band states at a high wave number ($k > 0$), called the Δ point. The Δ point is situated on the x axis (and on the equivalent y and z axes) near the end of this axis within the first Brillouin zone. This endpoint is called the X point. The corresponding wave number k_x is given by (diamond lattice)

$$k_x = 2\pi/a. \tag{8.1}$$

The Δ point in Si is roughly given by

$$k_\Delta = 4/5 k_x. \tag{8.2}$$

Transitions of carriers across the indirect bandgap need an energy transfer of E_{gind} and a momentum transfer of $\hbar k_\Delta$. The energy transfer is easily provided by a photon, whereas the large momentum transfer k_Δ needs the help of a phonon, which reduces the radiation transition probability of the indirect semiconductor. The wave number of the photon k_{phot} is rather small (about a factor 10^{-4}) compared to the Brillouin zone edge.

$$k_{\text{phot}} = \frac{q}{\hbar c} V_{\text{gind}} \tag{8.3}$$

Silicon-Based Photonics
Erich Kasper and Jinzhong Yu
Copyright © 2020 Jenny Stanford Publishing Pte. Ltd.
ISBN 978-981-4303-24-8 (Hardcover), 978-981-4303-25-5 (eBook)
www.jennystanford.com

168 | *Light Emitters*

The energy of the photon is usually measured by the corresponding voltage V_{gind} (this energy unit is called eV)

$$E_{\text{gind}} = qV_{\text{gind}} \tag{8.4}$$

The proportionality factor $q/\hbar c$ is equal to $5.06.10^6$/Vm. This means that a photon with around 1 eV energy has a negligible wave number compared to the wave number (about 10^{11}/m) of the Brillouin zone edge. Phonons have low energies (typically tens of millielectron volts), and they span in terms of wave number the full range of the Brillouin zone (for more details, see Chapter 2). Usually, photoluminescence (PL) from Si shows weak emission lines of energies below the bandgap (photon replica). The most significant contribution [1] stems from the transverse optical (TO) phonon, with energy of about 60 meV. Nevertheless, good radiative quantum efficiencies of about 10% may be obtained with very pure bulk Si and low injection levels. Then competitive nonradiative recombination channels (recombination from impurities and Auger recombination are weakly addressed) are also weak. Large emission volumes and low emission intensities prohibit usage in a device structure. The direct transition is much higher in energy than the indirect one. The lowest (Γ) point energy in a Si conduction band is more than 2 eV above the indirect (Δ) minimum. This energy difference for SiGe decreases but is always too high for a meaningful electron occupation probability except for pure Ge or GeSn.

Emission across the bandgap is the favored solution for semiconductor materials but not the only possibility. Let us consider two different paths of light generation without recombination across the bandgap. One is based on the incorporation of atoms with individual atomic emission properties, and the other relies on a nonequilibrium carrier distribution within the bands.

Rare earth elements have distinct energy level transitions that cover the full spectrum between near infrared and ultraviolet. The rare earth elements can be easily incorporated in glass. The optical stimulation of the emission is effective; also the electrical stimulation delivers good quantum efficiencies (10% and more), but the power efficiency [2] suffers from the high voltages (typically 100 V) needed to inject electrons into a thin glass layer. In semiconductors, the incorporation of rare earth elements is more difficult, resulting in rather low concentrations or poor crystal quality. This technique may be interesting for future photonic systems with integrated display functions.

A different mechanism of light emission gets active under strong electric fields easily realized in reverse-biased junctions. Electrons or holes are accelerated to get higher energies than under room temperature equilibrium. These carriers are accordingly called hot electrons (holes). Recombination of hot electrons with hot holes results in light emission [3] with energies larger than the bandgap energy. Finally, the hot carriers obtain energies (about 1.5 eV, or 40 times the equilibrium kinetic energy of an electron) to ionize neutral Si. Visible light with a broad spectral distribution is emitted from a junction biased toward avalanche breakdown, but breakdown is not a favorable mode of operation for a diode. Emission is more easily controlled with junctions that contain defects like dislocations. These junctions show a premature breakdown increase in current below the general breakdown voltage, which allows a safer operation. Defect engineering by thin, lattice mismatched SiGe layers yields diode structures with defined premature breakdown characteristics. This technique seems attractive for integration with standard complementary metal-oxide-semiconductor circuitry, and it could be used for a photonic monitoring system of safe operation of integrated circuits [4].

8.1 Bandgap Emission Mechanisms

First, we look at an ideal semiconductor with perfect periodicity. Then, we investigate the influence of structural modifications (superlattice periodicity; localization by porous Si and quantum dots) on the emission of indirect semiconductors.

8.1.1 Indirect Semiconductor Transitions

The principal structure of the bands of an indirect semiconductor is shown in Fig. 8.1. Carriers are injected, either optically or electrically, so that the band edges are populated with holes and electrons. We assume in the following that the quasi-Fermi energy levels F_p and F_n are within the bandgap. This allows approximating the carrier distributions within the band by the exponential Boltzmann equations. It is clear that the full Fermi–Dirac statistics have to be used for high injection when the quasi-Fermi levels lie within the bands, and we will mention this if necessary. The low-energy indirect bandgap is between the conduction band edge E_c and the valence band edge E_v (Fig. 8.1).

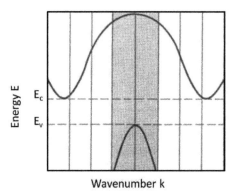

Figure 8.1 Schematic energy E diagrams versus wave number k of an indirect semiconductor (Si type with conduction band minimum at the Δ point in the k_x direction). The indirect bandgap E_{gind} is defined between the band edges E_c and E_v.

The main energy is released by a photon with energy $(E_g - E_{photon})$, where E_g is the bandgap energy and E_{photon} is the smaller phonon energy. The bandgap energy E_g measured by the light emission in slightly reduced by the excitation binding energy (excitonic bandgap). The exciton is a hole-electron pair that has to be created before a recombination process takes place. The exciton binding energy is rather small (Si: 10 meV; Ge: 3 meV). In doped material, the excitation is influenced by the charged dopant ions (bound exciton compared to free exciton in undoped material). These fine details can only be seen in low-temperature luminescence measurements. These details are overseen at higher temperatures (e.g., room temperature) by line brooding and by line shift (about $1/2k_BT$) from occupation statistics. Additionally, a weaker twin line at $E_g + E_{photon}$ appears that is caused by the capture of a phonon existing in the material at a higher temperature. The direct transition path is represented by a vertical line in the band diagram (Fig. 8.1) because of the negligible wave number contribution of phonons. The intensity is negligible in Si and SiGe but competes with the indirect path in Ge and GeSn, as explained in the following sections.

8.1.2 Brillouin Zone Folding from the Superlattice

The band structure of a semiconductor may be modified by an overlaid periodic potential that is realized by a repeated sequence of heterostructures. The basic idea dates back to the 1970s [5, 6]. An intuitive description of the modification of the band structure gives the zone following concept. The additional periodicity with the superlattice period length L shrinks the first Brillouin zone to $\pm \pi/L$. Let the length L be an integral multiple n of the periodicity length $a/2$ of the diamond lattice. (Remark: The periodicity length in the diamond lattice is $a/2$ because the cubic diamond lattice cell contains $8 = 2^3$ atoms). The first Brillouin zone of the superlattice is smaller by a factor n as the diamond lattice parent cell. It is created from the parent cell by consecutive folding (Fig. 8.2) and a horizontal bending at the zone edge.

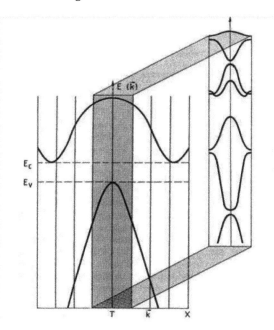

Figure 8.2 Reduction of the first Brillouin zone by a superlattice period (zone folding). Folding in four units is shown in the figure for the sake of clarity. Four minibands are created, which are separated by minigaps. The energy of the minigaps increases with a stronger superlattice potential. The essential effect for indirect semiconductors is given by the possibility to shift the position of the minimum of the conduction band with respect to the Brillouin zone edge.

By this construction, n minibands build from one parent band with small bandgaps at the zone edges. The small bandgaps at the zone edge increase with the strength of the superlattice potential. The superlattice is considered as a tailored artificial semiconductor with new electrical and optical properties. An indirect semiconductor material may change to a quasi-direct one by an appropriate choice of superlattice length [6]. First realizations with SiGe/Si superlattices started early [7], but confirmation of zone folding effects on the phonon spectra [8] and luminescence spectra [9] needed ultrathin Ge/Si superlattices with a period length L of 2.75 nm and 5.5 nm ($n = 5, 10, \ldots$). The transition strength [10] of the quasi-direct mode is 2–3 magnitudes lower than that of a direct semiconductor.

8.1.3 Localization of Wave Functions by Quantum Structures and Porous Si

The wave numbers of states in the infinite periodic semiconductor structure are well defined. Localization of wave functions results in uncertainties of the momentum, which relaxes the need for phonon contribution to the optical transition process. This can be already seen for luminescence from the random alloy SiGe, where the statistics distribution of the Ge atoms disturbs the perfect periodicity of the diamond lattice. The PL spectrum of SiGe quantum wells [11] is already dominated by the no-phonon emission line (Fig. 8.3) from the indirect bandgap.

The Heisenberg uncertainty principle says

$$\Delta x * \Delta p_x \geq \frac{h}{2} \tag{8.5}$$

A stricter localization is given by a quantization in all directions—a quantum dot (QD). Two effects appear under strong quantum confinement. One effect is bandgap widening because the confined states are higher in energy than the bottom of the well or dot. The other effect is delocalization of carriers in momentum space due to the uncertainty principle, making it possible for electrons and holes to recombine directly. Nanocrystalline silicon and porous silicon behave like a large assembly of QDs. Efficient luminescence can be obtained in the visible wavelength region.

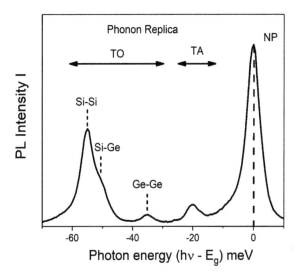

Figure 8.3 Photoluminescence intensity of SiGe layers versus photon energy ($hv - E_g$). The highest intensity of the random alloy provides the no-phonon peak (NP) at the excitonic bandgap. The phonon replica TO and TA (dominating in Si) are weaker and split up in three components, which can be identified as Si-Si, Si-Ge, and Ge-Ge vibrations.

8.2 Germanium Light-Emitting Diodes

For luminescence experiments, optical excitation at low temperatures is often applied (PL). For device applications, electrical excitation at room temperature is preferred (electroluminescence). The dominant electrical excitation uses diode structures for carrier injections (light-emitting diode [LED]).

8.2.1 Competition between Direct and Indirect Transitions

Ge is an indirect semiconductor like Si, but there are two essential differences: (i) the lowest lying conduction band has its minimum energy in the [111] direction at the edge of the first Brillouin zone (called the L point), and (ii) the energy difference (136 meV) between indirect and direct transition is much smaller than in Si. At room temperature, the direct bandgap is 0.80 eV and the indirect one is 0.664 eV.

The occupation probability of the direct states is much smaller than that of the indirect ones but not negligible like in Si. The direct occupation probability is about 0.4% of the indirect one within the Boltzmann approximation. Furthermore, the density of states of the direct conduction band is smaller than that of the indirect band (this is caused by the smaller effective mass of the electrons at the Γ point). Roughly, one can say that from 1000 electrons only 1 electron occupies the direct minimum; the other 999 are in the indirect minimum.

The transition probability of the direct bandgap is orders of magnitude higher than the indirect one. It is interesting to see which one is the more competitive. Early bulk Ge investigation with PL showed a strong indirect emission with a weaker direct shoulder [12]. Later investigation with epitaxial thin films [13] proved a stronger direct emission. The bulk results were explained [12] with the self-absorption of the direct line in thick substrates. The direct/indirect intensity ratio increases with temperature, favoring a room temperature operation of Ge LEDs. This observation is a consequence of the occupation statistics in indirect semiconductors, when higher temperature increases the direct/indirect occupation ratio.

A typical structure of a waveguide Ge LED is shown in Fig. 8.4 [14]. The vertical layer sequence starts with a silicon-on-insulator substrate.

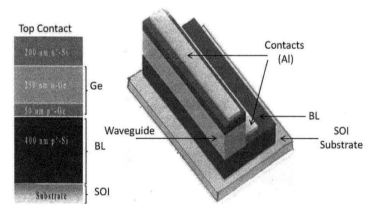

Figure 8.4 Typical layout and layer structure of a waveguide light emitter. The layer structure consists of a silicon-on-insulator (SOI) substrate with a highly p[+]-doped buried layer (BL) for the backside contact. The diode junction p/n is inside the Ge waveguide. The top contact layer is n[+]-Si, with Al as the top metal. Reprinted with permission from Ref. [14] © The Optical Society.

The silicon is p^+-doped to provide as the buried layer a good bottom contact property. Then follows a p^+-n Ge layer, which provides the diode structure for efficient forward current injections. The top layer is a thin n^+-Si, which provides good top contact properties and a well-behaved surface for the device technology steps.

The waveguide is structured by a dry etching step and metal contacts on the top (n-contact) and on the side (p-contact to the buried layer). The Ge LED emits light of about 1.5 μm wavelength into the waveguide when the contacts are forward biased. Interestingly, similar waveguide structures [15] can also be used for detection [16] and absorption modulation, which facilitates integration of active photonic devices.

8.2.2 Influence of Doping, Strain, and Sn Alloying

The chosen junction p^+/n is one sided and abrupt, and the junction extends to the n-side. The intensity of emission increases with the doping level at least up to an n-type doping of $3.10^{19}/cm^3$ [17]. The wavelength of emission increases by n-type doping, which is caused by the bandgap narrowing effect in highly doped semiconductors.

Another measure to increase the intensity and shift the wavelength is opened by applying tensile strain. Biaxial tensile stress [18] reduces the direct bandgap and decreases the energy difference between direct and indirect conduction band states. These effects increase the intensity and shift the wavelength. The small biaxial stress that is needed to cover the telecom bands around 1.55 μm is conventionally obtained by an annealing/cooling sequence. The tensile strain of up to 0.25% results from thermal mismatch in the expansion of the Si substrate and the Ge layer.

Alloying Ge with rather small amounts of Sn results in a reduction in the bandgap and a direct/indirect bandgap energy difference similar to the strain effect. Intensity increase and shift toward lower wavelengths are connected [19]. GeSn has a rather large lattice mismatch (14.6%) to Ge, and the equilibrium GeSn alloy is only thermodynamically stable up to rather small (<1%) Sn amounts. Metastable GeSn layers with higher amounts of Sn are grown at low temperatures [20]. Surface segregation of tin [21] and phase separation into a Ge-rich and tin-rich GeSn phase [22] limit the growth and processing conditions for device-related structures.

176 | *Light Emitters*

Stacked layers of GeSn and Ge [23] or GeSn and GeSiSn [24] are preferred multiquantum well structures for electrically stimulated LEDs. Research is ongoing to stabilize GeSiSn with high Sn amounts (>30%) on Si or Ge-on-Si platforms [25, 26].

References

1. V. Higgs, E. C. Lightowlers, G. Davies, F. Schäffler and E. Kasper (1989). Photoluminescence from MBE Si grown at low temperatures; Donor bound excitons and decorated dislocations, *Semicond. Sci. Technol.*, **4**, 593–598.

2. J. M. Sun, W. Skorupa, T. Dekorsy, M. Helm and L. Rebohle (2004). Efficient ultraviolet electroluminescence from a Gd-implanted silicon metal–oxide–semiconductor device, *Appl. Phys. Lett.*, **85**, 3387.

3. M. Morschbach, M. Oehme and E. Kasper (2007). Visible light emission by a reverse-biased integrated silicon diode, *IEEE Trans. Electron Devices*, **54**, 1091–1094.

4. E. Kasper and M. Morschbach (2013). Method for self-monitoring of breakdown in semiconductor components and semiconductor component constructed thereof, United States Patent US 8,519,732 B2.

5. L. Esaki and R. Tsu (1970). Superlattice and negative differential conductivity in semiconductors, *IBM J. Res. Dev.*, **14**, 61.

6. U. Gnutzmann and K. Clausecker (1974). Theory of direct optical transitions in an optical indirect semiconductor with a superlattice structure, *Appl. Phys.*, **3**, 9.

7. E. Kasper, H. Herzog and H. Kibbel (1975). A one-dimensional SiGe superlattice grown by UHV epitaxy, *Appl. Phys.*, **8**, 199–205.

8. H. Brugger, G. Abstreiter, H. Jorke, H. Herzog and E. Kasper (1986). Folded acoustic phonons in Si/SixGe1-x superlattices, *Phys. Rev.*, **B33**, 5928–5930.

9. R. Zachai, K. Eberl, G. Abstreiter, E. Kasper and H. Kibbel (1990). Photoluminescence in short-period Si/Ge strained layer superlattices, *Phys. Rev. Lett.*, **64**, 1055–1058.

10. G. Theodorou and E. Kasper (2007). Optical properties of Si/Ge superlattices, in *Landolt-Boernstein, New Series*, Vol. III, 34C3, E. Kasper and C. Klingshirn, eds. (Springer Verlag), pp. 50–86.

11. N. Usami and Y. Shiraki (2007). Single and coupled quantum wells: SiGe, in *Landolt-Boernstein, New Series*, Vol. III, 34C3 E. Kasper and C. Klingshirn, eds. (Springer Verlag), pp. 26–49.

12. T. Arguirov, M. Kittler, M. Oehme, N. V. Abrosimov, O. F. Vyvenko, E. Kasper and J. Schulze (2014). Luminescence from germanium and germanium on silicon, *Solid State Phenom.*, **205–206**, 383–393.

13. E. Kasper, M. Oehme, T. Arguirov, J. Werner, M. Kittler and J. Schulze (2012). Room temperature direct band gap emission from Ge p-i-n heterojunction photodiodes, *Adv. OptoElectron.*, **2012**, 916275.

14. R. Koerner, M. Oehme, M. Gollhofer, M. Schmid, K. Kostecki, S. Bechler, D. Widmann, E. Kasper and J. Schulze (2015). Electrically pumped lasing from Ge Fabry-Perot resonators on Si, *Opt. Express*, **23**, 14815–14822.

15. R. Koerner, M. Oehme, M. Gollhofer, M. Schmid, K. Kostecki, S. Bechler, D. Widmann, E. Kasper and J. Schulze (2015). Electrically pumped lasing from Ge Fabry–Perot resonators on Si, *Opt. Express*, **23**, 14815–14822.

16. A. Palmieri, M. Vallone, M. Calciati, A. Tibaldi, F. Bertazzi, G. Ghione and M. Goano (2018). Heterostructure modeling considerations for Ge-on-Si waveguide photodetectors, *Opt. Quantum Electron.*, **50**, 71.

17. M. Oehme, E. Kasper and J. Schulze (2013). GeSn heterojunction diode: detector and emitter in one device, *ECS J. Solid State Sci. Technol.*, **2**(4), 76–78.

18. M. Oehme, M. Gollhofer, D. Widmann, M. Schmid, M. Kaschel, E. Kasper and J. Schulze (2013). Direct bandgap narrowing in Ge LED's on Si substrates, *Opt. Express*, **21**(2), 2206–2211.

19. J. Michel, J. Liu and L. C. Kimerling (2010). High-performance Ge-on-Si photodetectors, *Nat. Photonics*, **4**, 527–533.

20. E. Kasper, M. Kittler, M. Oehme and T. Arguirov (2013). Germanium tin: silicon photonics toward the mid-infrared, *Photonics Res.*, **1**(2), 69–76.

21. E. Kasper (2016). Group IV heteroepitaxy on silicon for photonics, *J. Mater. Res.*, **31**, 3639–3648.

22. T. S. Perova, E. Kasper, M. Oehme, S. Cherevkov and J. Schulze (2017). Features of polarized Raman spectra for homogeneous and non-homogeneous compressively strained GeSn alloys, *J. Raman Spectrosc.*, **48**, 993–1001.

23. L. Kormos, M. Kratzer, K. Kostecki, M. Oehme, T. Sikola, E. Kasper, J. Schulze and C. Teichert (2017). Surface analysis of epitaxially grown GeSn alloys with Sn contents between 15% and 18%, *Surf. Interface Anal.*, **49**, 297–302.

24. B. Schwartz, P. Saring, T. Arguirov, M. Oehme, E. Kasper, J. Schulze and M. Kittler (2016). Analysis of EL emitted by LEDs on Si substrates

containing GeSn/Ge multi quantum wells as active layers, *Solid State Phenom.*, **242**, 361–367.

25. N. von den Driesch, et al. (2018). Advanced GeSn/SiGeSn group IV heterostructure lasers, *Adv. Sci.*, **5**, 1700955.

26. V. A. Timofeev, A. I. Nikiforov, A. R. Tuktamyshev, V. I. Mashanov, I. D. Loshkarev, A. A. Bloshkin and A. K. Gutakovskii (2018). Pseudomorphic GeSiSn, SiSn and Ge layers in strained heterostructures, *Nanotechnology*, **29**, 154002.

27. W. Dou, M. Benamara, A. Mosleh, J. Margetis, P. Grant, Y. Zhou, S. Al-Kabi, W. Du, J. Tolle, B. Li, M. Mortazavi and S.-Q. Yu (2018). Investigation of GeSn strain relaxation and spontaneous composition gradient for low - defect and high-Sn alloy growth, *Sci. Rep.*, **8**, 5640.

Chapter 9

Detectors

For a chosen wavelength of light, the material demands for waveguides and detector/light sources are contradicting with respect to matter-light interaction. For the waveguide one wants absorption as low as possible, and for the detector it should be high. That means the material for detectors has to be different from that of the waveguides, leading to two prevalent system concepts for Si photonics:

- High-bandgap (insulator) waveguide with a Si detector
- Si waveguide with a low-bandgap detector

In this chapter, we discuss detector principles, system and wavelength considerations, detector structures, and high-speed operation and give selected experimental results.

9.1 Detection Principles

In principle, each matter-light interaction could be used to sense impinging light. In practice three main principles (Table 9.1) dominate: photon detectors, thermal detectors, and coherent detection (heterodyne detection).

Silicon-Based Photonics
Erich Kasper and Jinzhong Yu
Copyright © 2020 Jenny Stanford Publishing Pte. Ltd.
ISBN 978-981-4303-24-8 (Hardcover), 978-981-4303-25-5 (eBook)
www.jennystanford.com

180 | *Detectors*

Table 9.1 Detection principles used in photodetectors

Photon detectors	Thermal detectors	Coherent detection (amplitude/phase)
Chemical change > photoplates Photoconductors (intrinsic/ extrinsic) Photodiodes/phototransistors Photoemissive detectors (photomultipliers, microchannel plates)	Bolometers	Mixing with a coherent local oscillator/laser

The broadest group—photon detectors—uses the large variety of effects when light meets atoms or solids. They create chemical changes in photoresists or photoplates (the original effect in photography), modulate the conductivity in photoconductors, or generate photocurrents in photodiodes/phototransistors and solar cells, or a different frequency is generated by mixing with a local oscillator/laser in coherent detection (a method that is the common technical choice in high-performance electrical detection—but difficulties with laser integration and phased lock-loop frequency stabilization prevented their rapid introduction to optoelectronic detection). The heating caused by the absorbed energy is used in bolometers for mid- and far-infrared (MIR and FIR) detectors where the low photon energy requires cooling of small-bandgap semiconductors in photon detectors. In a semiconductor system otherwise photocurrent generation in photodiodes/phototransistors is the preferred method for light detection. In phototransistors the detecting part is also a diode function; the transistor structure may be used for first-stage amplification.

The discussion for a certain detector is mainly made with respect to the photodetector parameters given in Table 9.2.

Table 9.2 Important parameters for photodetectors

Quantum efficiency	Spectral response
Noise bandwidth	Time response/frequency
Linearity	Size and numbers of pixels
Dynamic range	Operation temperature

For integrated systems the spectral response, the quantum efficiency, and the frequency bandwidth are the most important parameters.

9.2 Detector Configuration and Wavelength Considerations

Of primary importance for the detector configuration is the angle of incidence of light.

9.2.1 Vertical Incidence Detection

For image sensors a vertical incidence is the standard situation. Figure 9.1 shows the scheme of a future detector array embedded in a system on chip (SOC) consisting of the detector array, a logic part in complementary metal-oxide silicon (CMOS) technology, and high-frequency (radio frequency [RF]) input/output connections. Heterobipolar transistors (HBTs) for millimeter-wave operation speed (30–300 GHz) drive the RF components.

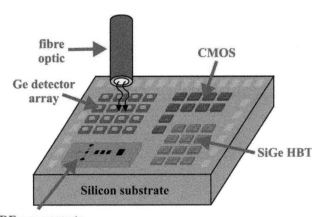

Figure 9.1 Integrated Ge detector array in a system on chip (SOC) comprising the detector array, a high frequency (RF) part with SiGe HBTs, and a CMOS logic. Reprinted by permission from Springer Nature Customer Service Centre GmbH: Springer Nature, *Frontiers of Optoelectronics in China*, Ref. [1], copyright (2010).

In this figure, the array is made from Ge detectors for near-infrared (NIR) vision as requested in automotive assistance systems for pedestrian detection. Vertical incidence allows efficient and fast detection if the absorption is strong at the selected wavelength. In detector arrays, a read-out circuitry is added to the sensor. The scheme is shown (Fig. 9.2) in the example of the CMOS image sensors, which are now in broad use in consumer electronics.

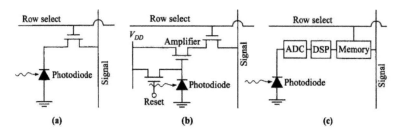

Figure 9.2 Different versions of CMOS image sensors: (a) passive pixel sensor (PPS); (b) active pixel sensor (APS); and (c) digital pixel sensor (DPS).

The simple form is the passive pixel sensor (Fig. 9.2a), where the photodiode is connected with a metal-oxide-silicon (MOS) transistor whose gate is terminated at the row line whereas the signal line is in contact with the transistor output (drain). With a proper gate voltage on the row select line all transistors of this line are on and the signals of the photodiodes on this row line can be read out with the signal lines. Active pixel sensors have already included an amplifier to read out the amplified signal (Fig. 9.2b). Digital pixel sensors give their analog signal to an analog-digital converter, where it is converted to a digital signal, processed in a digital signal processor, and saved in a memory.

9.2.2 Lateral Incidence Detection

A lateral incidence is more convenient for on-chip waveguide systems and for coupling to fiber optics. Even for original vertical incidence on the chip, a coupling structure to lateral incidence on the photodiode is applied if relaxed adjustment and high-speed operation are in demand. We will discuss this topic in more detail in the speed section. A scheme of lateral incidence of light via a planar waveguide (right side) is shown in Fig. 9.3, where the high-frequency electrical contacts (left side) are already given as **co**planar electrical **w**aveguide (CPW) in a ground (G)–signal (S)–ground (G) configuration. CPWs from thin metal lines allow electrical signal transfer at high speeds of up to more than 100 GHz.

(Remark: Do not confuse the abbreviations G and S used for ground and signal contacts with the same abbreviations used for gate and source contacts in transistor technology).

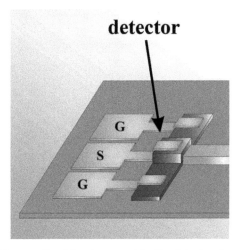

Figure 9.3 Scheme of a lateral waveguide Ge detector (arrow) with light coupled in from a Si waveguide. The Si waveguide (coming from the right side) is butt-coupled to the Ge detector. The high-speed electrical signal readout (on the left side) is realized by a coplanar electrical waveguide (CPW) with symmetric ground contacts (G) to the bottom layer and the central signal line (S) to the top layer of the Ge-on-Si p-i-n detector structure.

9.2.3 Wavelength Selection

Either the selected wavelength is defined by the needs of an intended application, or it is selected by a tradeoff between properties of available materials, technologies, and devices. Some applications need specific wavelengths; for example, processing of telecommunication signals is concentrated on bands around 1.3 μm and 1.55 μm. Sensors for environmental gas detection, for example, for methane or carbon dioxide, rely on larger wavelengths in the MIR regime because there are specific absorption bands for molecules.

The wavelengths of on-chip data transfer may be chosen to fulfill high bandwidths and convenient technological device fabrication. Si waveguides are transparent for wavelengths above 1.2 μm. Foundry service for Ge on Si devices is available that opens the wavelength window up to 1.6 µm for slightly tensile strained Ge. Devices with cutoff wavelengths above 1.6 µm are available in cooperation with research institutions focusing on highly tensile strained Ge, metastable GeSiSn alloys, or III/V epitaxy on Si.

For the detector material, a rather small absorption depth d_α is requested to minimize device size and to improve device speed.

$$d_\alpha = 1/\alpha \qquad (9.1)$$

Practical values of α are typically in the range of 10^3/cm to 10^5/cm, which corresponds to absorption depths from 10 µm to 0.1 µm. Exceptions from this rule are allowed in silicon, where the high quality of material delivers diffusion lengths of up to millimeters, facilitating minority carrier transport from even the substrate backside to the frontside junction. This can be exploited in solar cells or slow photodetectors where an absorption range from $\alpha = 10^{-1}$/cm can be utilized.

In the following, we consider the absorption properties of some specific semiconductors (for a general discussion of absorption see Chapter 3), with emphasis on the SiGe material system.

In Fig. 9.4, the absorption coefficient α is given for several common semiconductors: GaAs, Si, Ge, and GaInAs. The III/V

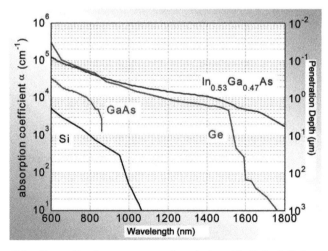

Figure 9.4 Absorption in common semiconductors (Si, GaAs, InGaAs, and Ge are selected). Given is the absorption coefficient as a function of wavelength.

Detector Configuration and Wavelength Considerations | **185**

semiconductors GaAs and GaInAs are direct semiconductors with different bandgaps. Above the bandgap energy E_{gdir}, photons are easily absorbed—in the often-chosen presentation as a function of wavelength (like in Fig. 9.4) this means strong absorption below the corresponding wavelength λ_g, where

$$\lambda_g \,(\mu m) = 1.24 \,\mu m/E_g \,(eV). \tag{9.2}$$

Both silicon and germanium are indirect semiconductors [2] where absorption below the wavelength λ_g increases much slower, as shown in the Si case. Fundamental absorption [3] starts below λ_g = 1.1 µm but increases only up to $5 \cdot 10^3$/cm at λ = 0.6 µm, nearly an order of magnitude below GaAs, for which absorption starts at λ_g = 0.93 µm, steeply crossing the Si absorption curve at about λ = 0.85 µm. However, hybrid absorption curve characteristics are seen if one looks at Ge. At the bandgap wavelength (λ = 1.85 µm) the absorption curve starts slowly, like in Si, but suddenly at about λ = 1.6 µm, the curve steepens and reaches that of the direct semiconductor InGaAs. The explanation for these absorption characteristics of Ge is given by its conduction band details [4]. For the explanation, let us start with the more familiar conduction band of Si (Fig. 9.5).

The bandgap is defined by the electron energy difference between the conduction band minimum and the valence band maximum. The valence band maximum is always at the wave vector length k = 0 (named the Γ point), whereas the electron energy minimum of the conduction band is at a finite k value in the case of an indirect semiconductor. In the case of Si, the minimum conduction band energy is at the Δ (delta) point of the Brillouin zone, which is in the x direction (in Si 0.8X). In the cubic diamond lattice, the y and z directions and their minus signs are equal to the x direction, which means that the energy minimum is sixfold degenerate, as best seen in a constant energy surface (see Fig. 9.5, top) presentation. The lowest direct transition is far above this indirect bandgap energy, at about E_{gdir} = 3.4e V, which corresponds to an ultraviolet (UV) photon (λ = 364 nm). In Si, the absorption in the NIR and the visible (VIS) spectrum is solely dominated by the indirect absorption. In Ge the lowest direct transition comes strongly down to E_{gdir} = 0.80 eV whereas the X transition is only slightly reduced compared to Si (E_{gX} = 0.85 eV). In spite of this crossing of Γ and X transition the semiconductor Ge is an indirect one because the L transitions (111 direction, cube diagonal) decrease more than the X transitions to E_{gL} = 0.66 eV (all values given for room temperature). The degeneracy of the L minimum in the conduction band is fourfold; see the constant energy surface in Fig. 9.5 (eight energy ellipsoids are drawn, but energy ellipsoids at the zone edge between first and second Brillouin zones count only half for the measure of degeneracy). This indirect bandgap causes the weak absorption start at $\lambda = 1.85$ μm whereas the only slightly higher direct transition ($\lambda = 1.55$ μm) is responsible for the steep absorption increase around that direct transition. Heating the device (self-heating or forced heating) reduces the bandgap of diamond-type semiconductors and extends by that way the infrared range.

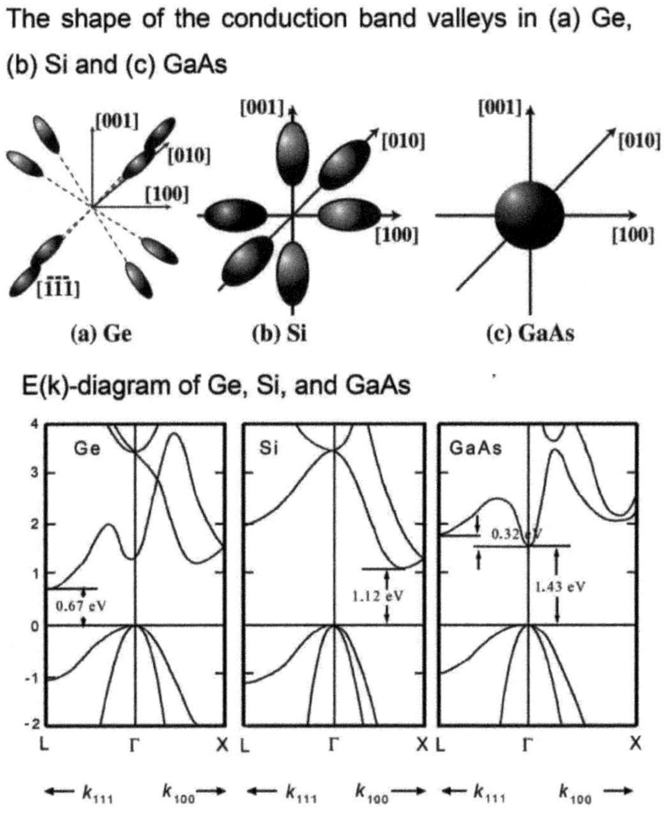
The shape of the conduction band valleys in (a) Ge, (b) Si and (c) GaAs
[001]
[010]
[100]
[111]
(a) Ge
(b) Si
(c) GaAs
E(k)-diagram of Ge, Si, and GaAs
Ge
Si
GaAs
0.67 eV
1.12 eV
1.43 eV
L
Γ
X
k111
k100

Figure 9.5 Constant energy surfaces for conduction band electrons (top part of the figure). Band structure (energy E versus wave vector length k) of common semiconductors Si, GaAs, and Ge (lower part of the figure). Given are the energies of carriers as a function of the wave vector k (center Γ, $k = 0$, X point in 100 direction, L point in 111 direction), using data from Refs. [1, 3, 4].

The transition from Si to Ge can easily be observed with SiGe alloys with increasing Ge amounts (Fig. 9.6).

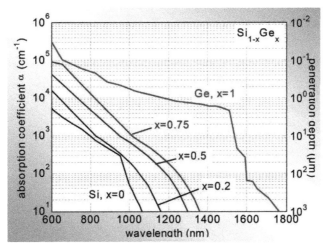

Figure 9.6 Absorption versus wavelength of SiGe alloys with different Ge amounts x, from pure Si ($x = 0$) to pure Ge ($x = 1$).

In Fig. 9.6, SiGe alloys with Ge amounts $x = 0, 0.2, 0.5, 0.75$, and 1 are selected. With an increasing Ge content the absorption curve is shifted to the infrared corresponding to the lowering of the indirect bandgap energy. The bandgap lowering ΔE_{gx} of SiGe (compared to Si) is nonlinear, given in parabolic description [2] as

$$\Delta E_{gx} = -0.43x + 0.206x^2 \text{ (in eV) for } x \leq 0.85. \quad (9.3)$$

The Δ valley is the lowest one up to a Ge amount $x = 0.85$. All SiGe alloys up to very high Ge amounts (85%) have a Si-like conduction band minimum (Δ point). Only above $x \geq 0.85$ the L minimum comes down below the Δ minimum. The L minimum in the Ge-rich range ($x > 0.85$) is described by a linear dependence on x. We compare now the indirect L minimum energy of SiGe with that of Ge:

$$\Delta E_{gL} \text{ (compared to Ge)} = 1.27(1 - x) \quad \text{(in eV)} \quad (9.4)$$

An extrapolation of this linear behavior yields an L minimum of 2.01 eV for Si, which is within the reasonable range of what is shown in Fig. 9.5.

The bandgaps deduced from photoluminescence (PL) and absorption differ by up to 40 meV. A part of the difference may be

188 | *Detectors*

explained by the finite exciton (electron-hole pair) binding energy E_b, which varies from 14.5 meV for Si excitons to 4.1 meV for Ge excitons. From PL measurements, the exciton bandgap $(E_g - E_b)$ will be measured at low temperatures. With increasing temperature, the bandgap E_g shrinks, for example, for Si from 1.17 eV at 0 K to 1.118 eV at 300 K or for SiGe ($x = 0.78$) from 0.915 eV to 0.87 eV. This means the Δ minimum (remember this minimum dominates up to $x = 0.85$) shrinks at about 5% for 300 K independent of the Ge content. The temperature dependence of the L minimum is definitely stronger, at about 11% for 300 K in Ge (0.74 eV at 0 K; 0.66 eV at 300 K).

Within the discrepancy of 40 meV between emission and absorption experiments, the bandgap of unstrained SiGe is well described. What is the situation with the direct transitions for the whole range of SiGe compositions? The critical points of the density of states (named E_0, E_1, and E_2) are mainly investigated by spectroscopic ellipsometry, reflectometry, and modulation spectroscopy (electroreflectance and thermoreflectance). The most important signatures in refractive index and absorption stem from the E_1 and E_2 transitions. The E_1 transition marks the highest n value, whereas the absorption peak is near the E_2 transition [3]. The E_1 transition shifts with the Ge amount x as follows:

$$E_1(x) = 3.395 - 1.44x + 0.153x^2 \text{ (in eV)} \qquad (9.5)$$

From ellipsometry or reflectometry the Ge content may be deduced from the E_1 position

$$x = 4.707 - (6.538E_1 - 0.0397)^{0.5}. \qquad (9.6)$$

The spin-orbit split energy gap $E_1 + \Delta_1$ is slightly larger:

$$E_1 + \Delta_1 = 3.428 - 1.294x + 0.062x^2 \text{ (eV)} \qquad (9.7)$$

The split Δ_1 increases from 33 meV at Si to 88 meV at the Ge side. The E_2 transition (near the X point) is nearly independent of the Ge amount, $E_2 = 4.4$ eV. The direct transitions at the Γ point ($k = 0$) are E_0, $E_0 + \Delta_0$, and E_0'. The gap E_0' decreases very slightly with Ge amount

$$E_0' = 3.40 - 0.3x \qquad \text{(in eV).} \qquad (9.8)$$

The transition, which dramatically changes with Ge amount, is the E_0 transition.

This transition comes down from 4.05 eV in Si to 0.8 eV in Ge, and it is the lowest direct transition up from about $x = 0.35$ (on the Si side the E_0' and E_1 transitions are the lowest direct ones).

Linear interpolation gives

$$E_0 = 4.05 - 3.25x \text{ (in eV)}. \tag{9.9}$$

The spin-orbit split energy Δ_0 increases with the Ge amount from 44 meV to 290 meV.

The strong dependence of E_0 transition on the Ge amount causes Ge to be a pseudo-direct absorber because only a further shift of 140 meV would be necessary to cross the direct Γ and indirect L transitions. In emission, the indirect character dominates in relaxed Ge, but we will discuss in the following chapters strategies with strain, alloying, and doping to also strengthen direct emission processes.

In Fig. 9.7 we summarize the results for unstrained SiGe alloys.

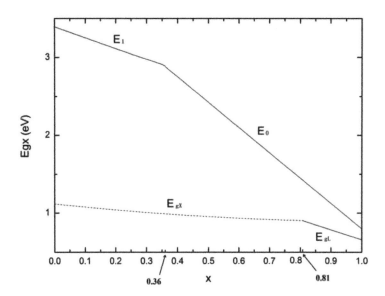

Figure 9.7 Direct and indirect band transitions in SiGe alloys. Shown are only the lowest direct transitions (E_1 at a low Ge amount and E_0 at a higher amount) and the lowest indirect transitions E_{gX} and E_{gL}.

The solid line shows the lowest direct transition (E_1 in SiGe, $x \leq 0.35$; E_0 in SiGe with $x > 0.35$) compared with the dotted line for the lowest indirect transition (X transition, $x < 0.85$; L transition, $x \geq 0.85$). On the Si side the direct transitions E_0' and E_1 need about

3 eV (UV light necessary) whereas the indirect transition X needs only more than 1.1 eV. Above $x = 0.35$ the strongly decreasing E_0 gap defines the lowest direct transition whereas above $x = 0.85$ the indirect L gap replaces the X gap.

We have already discussed the temperature dependence of indirect transitions. The temperature coefficients of the direct transitions are found to be on the order of -0.2 meV/K to -0.5 meV/K. The main reason for the negative temperature effects is given by the lattice expansion, which is accompanied by shrinking gaps.

The majority of data of unstrained alloys are collected from bulk samples. In heterostructure layers both compressive and tensile strain are usually found as a result of lattice mismatch (SiGe on Si is compressively strained) and thermal expansion mismatch (unstrained Ge/Si cooled down from epitaxy temperatures to room temperature exhibits tensile strain). For heterostructure devices the strain status has to be defined and the strain effects have to be described and explained.

9.3 Photon Detector Structure

A photoconductive device realizes the simplest photodetector. It exploits the change in conductivity of a semiconductor that is illuminated. The intrinsic photoconductivity involves the excitation of electrons and holes from a photon absorption process. This process occurs when the energy of the photon exceeds the bandgap energy. The technological realization needs ohmic contacts on both sides of the illuminated area. A voltage bias is applied on these contacts. The current through the device increases under illumination because of the additional carriers caused by absorption. This detector principle functions not only for a single crystalline material but also for polycrystalline or amorphous materials. Therefore, it is frequently used in early phases of material testing when technology is under development.

Photodiode detectors dominate in the VIS and NIR spectrum range because of good properties in detectivity and speed. In the following sections, we will discuss different technological realizations of photodiode detectors.

9.3.1 P/N Junction Photodiode

The basic detector structure [4] is shown in the example of a p-i-n diode (Fig. 9.8).

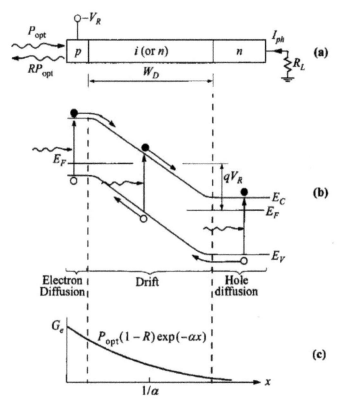

Figure 9.8 Operation of a p/n junction photodiode: (a) Cross-sectional view of the p-i-n diode. (b) Energy band diagram under a reverse bias. (c) Carrier generation characteristics (using data from Refs. [1, 4]).

A nominal intrinsic semiconductor (i-layer) absorption layer (Fig. 9.8a) is terminated on both sides by highly doped layers of the opposite carrier type. In Fig. 9.8a, light of power P_{opt} is entering from the left. A part $R \cdot P_{opt}$ is reflected (R is reflectivity) so that the power $P_{opt} \cdot (1 - R)$ penetrates into the semiconductor. On a bare semiconductor the reflectivity R is high because of the high refractive index n of semiconductor materials.

$$R = \left(\frac{n-1}{n+1}\right)^2 \tag{9.10}$$

In the very high absorption spectral range, the complex n value has to be taken, which then reads

$$R = 1 - 4n/(n^2 + \kappa^2 + 2n + 1). \tag{9.11}$$

The relative change in reflectivity

$$\frac{dR}{dk} = \frac{(1-R)^2}{2n} \cdot \kappa \tag{9.12}$$

is proportional to the absorption index κ but with a low prefactor of about 0.05 for a semiconductor with a refractive index $n = 4$. A considerable shift in R (>10%) by the influence of the very high absorption index κ is given roughly by a κ value of more than 1. Looking at a table of optical constants [3], one finds for Si, SiGe, and Ge an upper photon energy below which absorption is nearly negligible for the calculation of reflectivity.

These energy levels are 3.25, 3.1, 3, 2.85, 2.6, 2.3, and 2 eV for the above given SiGe alloy row with $x = 0, 0.1, 0.2, 0.3, 0.5, 0.75$, and 1, respectively.

The high reflectivity of semiconductors makes antireflection coatings of photodiodes a must for exploitation of the full responsivity potential. We will discuss technical details in the next section.

The electric field $\bar{F}$ in an ideal intrinsic (i) layer is easily calculated by

$$\bar{F}_i = \text{const.} = (V_{bi} - V)/d_i, \tag{9.13}$$

with V_{bi}, V, and d_i as built-in voltage, applied voltage (negative value for reverse diode voltage!), and thickness of the intrinsic layer, respectively. The field is only slightly penetrating in the surrounding highly doped layers. Let us assume for this first approach that the doping in the n and p layers is high enough for one to be able to neglect this field penetration. The energy band diagram is then as shown for a reverse bias in Fig. 9.8b. The energy is linearly varying if the field is constant. Figure 9.8c shows the spatial distribution of the carrier generation rate $G_e = G_p = G$ inside the semiconductor when for simplicity no backside reflection is considered.

$$G = (1-R)\exp(-\alpha x) \cdot P_{opt} / A \cdot (\hbar\omega) \tag{9.14}$$

Here R is reflectivity, α is the absorption coefficient, P_{opt}/A is the power density (A is area), and $\hbar\omega$ is photon energy.

A note on the term intrinsic is necessary before we discuss the electrical field $\bar{F}$ and the band diagram. Physically correct, the term "intrinsic" would require that the doping be below the intrinsic carrier concentration. In silicon the room temperature intrinsic carrier concentration is about 10^{10} cm^{-3}, which is orders of magnitude below the technical possibilities of 10^{13} cm^{-3} – 10^{16} cm^{-3} background doping N_b in bulk or epitaxial layers. In a technical sense, the term "intrinsic" is less strictly used when the i-layer is fully depleted at zero bias (at the built-in voltage V_{bi}). The i-layer is termed ν-layer (n-background) or π-layer (p-background), respectively, if the type of background doping is known. The electric field decreases linearly from a maximum at the junction (n$^+$/p or p$^+$/n for p- or n-background, respectively). The slope of the field strength decrease depends on the background doping.

A one-sided abrupt p/n junction follows immediately from the p-i-n if the intrinsic layer is higher doped (e.g., in the 10^{17} cm^{-3} range compared to the 10^{20} cm^{-3} range for the electrode layers). Then the depletion layer ends within the now-doped absorption layer.

The following summarizes some textbook formulas for the basic properties of the depletion layer in a one-sided p/n junction (p-doping high and n-doping moderate with density N_D).

The built-in voltage

$$V_{bi} = (k_B T / q)\ln(N_D \cdot N_A / n_i^2). \tag{9.15}$$

Here N_A is doping of the p electrode and n_i is intrinsic carrier density.

The depletion width

$$d_{depl}^2 = (2\varepsilon_S / q)(1 / N_D)(V_{bi} - V). \tag{9.16}$$

Here ε_S is the dielectric constant of the semiconductor and V the applied voltage (positive sign in forward direction). The maximum field strength E_m (at the interface) is as follows:

$$E_m^2 = (2q / \varepsilon_S) \cdot N_D \cdot (V_{bi} - V) \tag{9.17}$$

or

$$E_m = \left(\frac{q}{\varepsilon_S}\right) N_D \cdot d_{depl} \tag{9.18}$$

(if d_{depl} is known).

194 | *Detectors*

These textbook formulas use the so-called Schottky approximation, which assumes charge contributions only from the dopant atoms in the depletion region.

Approximations that are more refined take care of the diffusion tails of the depletion edge and of bandgap shrinkage in highly doped layers. The diffusion tails lead to a slight modification of Eq. 9.15 in that V_{bi} is replaced by $(V_{bi} - 2k_BT/q)$; the high doping causes a bandgap shrinkage, which is described in electronic devices by an increase in the intrinsic carrier concentration. Furthermore, the full Fermi–Dirac statistics have to be used instead of the Boltzmann approximation. For details, the interested reader should refer to books on semiconductor device physics [5].

The generation rate G decreases, like light intensity, exponentially with the absorption penetration depth as decay length. We now have to determine whether a carrier pair is generated in the depletion region (i-region) or in the surrounding region. A pair generated in the depletion region is separated by the high electric field, and the carriers are collected by the p and n electrodes if the quality of the material is not too bad. The pairs generated in the surrounding regions move randomly because of the lack of an electric field. The minority carriers may diffuse to the junction and contribute to the photocurrent. The diffusion is strong within the diffusion lengths L_n and L_p for electrons (in p material) and holes (in n materials), respectively. The diffusion length L is connected to two fundamental properties of a semiconductor, the minority carrier diffusion coefficient D and the recombination lifetime τ. For electrons the relation reads (for holes the index is p)

$$L_n^2 = D_n \cdot \tau_n .\qquad(9.19)$$

This is a general law for Brownian motion linking the square of the length with the product of the diffusion constant times the time. The diffusion constants are linked to the mobility μ via

$$D_n = \mu_n (k_B T / q).\qquad(9.20)$$

Here the thermal voltage $k_B T/q$ is the proportionality constant (k_B is Boltzmann constant, T is temperature, and q is the electron charge).

The diffusion length L is rather large in low-doped semiconductors, and there it is used as a quality criterion because the recombination lifetime τ in indirect semiconductors is given by the recombination across midgap energy levels of metallic or defect traps. In high-doped semiconductors the lifetime is strongly reduced by Auger recombination (transfer of the momentum to a nearby carrier instead of phonon generation), resulting in submicrometer diffusion lengths in the p and n electrodes. Diffusion increases the internal quantum efficiency in solar cells, for which quantum efficiency is more important than speed (a field-free motion like diffusion is much slower compared to the fast velocity of electrons/holes in the strong electric field of the depletion layer). In high-speed applications, the diffusion current is minimized despite design constraints to get high quantum efficiencies. The design constraints of microwave and millimeter-wave-frequency-modulated detectors are treated in a separate section later. Here, only an order of magnitude assessment of the time constants in diffusion is given. Assume $L = 100$ nm and $D = 1$ cm^2/s; then the time constant will be 0.1 ns, which is acceptable up to a modulation frequency of 1 GHz.

The given relations are valid not only for Si but also for other semiconductors. The numerical values of the depletion layer properties differ because of different values of the intrinsic carrier density n_i. In heterojunctions, the band discontinuities disrupt the smooth energy function known from homojunctions. The band discontinuities of the valence and conduction bands are a typical electronic property of a pair of semiconductors. They depend on the chemical composition and the strain status of the heterostructure couple.

The key mechanism of all photocurrent-based detectors is the separation of carriers by the electric field of a depletion layer. In different types of detectors this depletion layer may be created instead of a p-i-n or p/n junction, by metal-semiconductor junctions (Schottky contact) or metal-insulator-semiconductor (MIS) junctions (MIS or MOS diode).

9.3.2 Schottky Photodiode

In a metal-semiconductor junction (Schottky contact), the depletion layer is caused by the charge transfer due to the work function differences of metal and semiconductor. For the depletion layer formula only a lower built-in voltage appears, replacing Eq. 9.15 by Eq. 9.21.

$$V_{bi} = (Schottky - contact) = \Phi_B - \Phi_{nF} \qquad (9.21)$$

Φ_B is the Schottky barrier height of the metal/semiconductor pair, which in the Anderson model without interface charges is given by the difference between work function of the metal and electron affinity of the semiconductor. Φ_{nF} is the potential difference between conduction band and Fermi level. The device is called a metal-semiconductor-metal (MSM) diode because the backside contact is an ohmic metal contact. MSM photodiodes have the advantage of easy technological realization but suffer from high dark currents. The high dark currents are caused by the lower built-voltage V_{bi} of Schottky contacts compared to p/n junctions, for example, in Si, 0.3–0.6V versus 0.8–1 V. To solve the dark current problem, a thin insulating layer is placed between the metal and the semiconductor. This junction is called MIS, or it is called MOS if the insulator is silicon oxide. The insulating oxide separates the depletion layer from the gate metal contact. We have the same work function difference like in an MSM diode, but now part of the potential difference drops across the insulator (V_i). The potential difference along the depletion layer is now governed by the surface potential Ψ_s, which replaces ($V_{bi} - V$) given for p/n and Schottky photodiodes in Eqs. 9.15–9.20.

Consider a p-type substrate without interface charges. The work function difference is described by the flat band voltage V_{FB}. The surface potential Ψ_s is given by

$$\Psi_s = -V_{FB} + V - V_i, \tag{9.22}$$

where

$$V_i = |Q_s| d_i / \varepsilon_i. \tag{9.23}$$

Here Q_s is the total charge in the semiconductor depletion layer and d_i and ε_i are the thickness and the dielectric constant (permittivity), respectively, of the insulator. Ideally, the flat band voltage V_{FB} is defined by the work function difference between metal and semiconductor. Corrections have to be made in the case of oxide charges. The voltage drop across the insulator is given by V_i. The depletion layer into the semiconductor is described by the same equations (Eqs. 9.17 and 9.18) as before but the potential difference ($V_{bi} - V$) in the p/n junction is now replaced by the surface potential Ψ_s. In MOS varactor or transistor operation two additional modes are available, inversion and accumulation. A positive gate voltage higher than the so-called threshold voltage V_{th} (in our example of a p substrate) induces a thin n-layer—the n-inversion channel—below the oxide. A negative voltage below the flat band voltage V_{FB} removes the depletion layer and induces a majority carrier increase (p accumulation) of the oxide interface. The conventional n-channel MOS transistor operates between inversion (on: the n channel is connected to the n-doped source + drain regions) and the depletion (off). The gate insulator should be thick enough (silicon oxide thicker than about 3 nm) to avoid significant tunneling currents. The MIS photodiode operates differently. There are no connections to source/drain regions, and the insulator is thin enough to allow tunneling of carriers. Figure 9.9 shows a typical arrangement for MIS photodiodes for vertical light incidence [6]. For simplicity, the backside contact is shown on the substrate backside. In integrated circuits, the backside contact is placed on the front side and connected to the device by a highly doped buried layer.

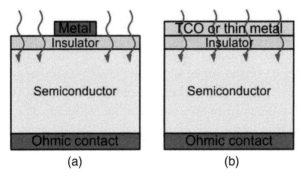

Figure 9.9 The schematic structures of MIS photodetectors [6]. (a) Light radiates into the semiconductor from the unshadowed region when the metal is not transparent. (b) Light radiates into the semiconductor passing through the transparent conducting oxide (TCO) or thin semitransparent metal film. Reproduced under open access (CC BY) from Ref. [6] (https://www.mdpi.com/openaccess).

A normal metal gate shadows the incident vertical light (Fig. 9.9a) so that absorption takes place only around the metal finger. A preferred embodiment of the MIS uses transparent gate electrodes (Fig. 9.9b), either a thin semitransparent metal film or a transparent conducting oxide. The current mechanisms of an MIS diode are shown in Fig. 9.10. An inversion bias (positive for a p substrate) is applied. In Fig. 9.10a, the dark current is shown.

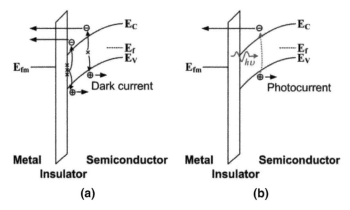

Figure 9.10 The current mechanisms of an MIS diode [6]. (a) Without light incidence; (b) with light incidence. A positive bias (inversion bias) is applied for this metal/insulator/p-type semiconductor example. Reproduced under open access (CC BY) from Ref. [6] (https://www.mdpi.com/openaccess).

Although an inversion bias ($V > V_{th}$) is applied, no inversion layer is created in the MIS photodiode! This is because the inversion channel in an MIS diode is built up by minority carriers generated in the depletion region. In the MIS photodiode these minority carriers leak through tunneling to the gate and define the dark current of the device (Note: In an MOS transistor the inversion channel is fed by the highly doped source/drain regions. This explains the much higher dark current in transistors with thin, leaky insulators). Absorption of light (Fig. 9.10b) adds hole-electron pairs that are separated by the electric field of the depletion layer and measured as photocurrent. Forward bias shifts, however, the metal Fermi level above the semiconductor conduction band, which results in strong electron tunneling from the metal side to the semiconductor, delivering a high current with near-ohmic contact properties. This operation mode is used in interdigitated contact fingers to create both contacts of the photodiode at the front side of the chip (metal-insulator-metal).

9.3.3 Avalanche Photodetector

An integrated or external low noise amplifier, as discussed with the active CMOS pixel, may amplify the photocurrent. Alternatively, an internal amplification mechanism may be sought. The most important physical mechanism for internal amplification is impact ionization in a strong electric field, which causes avalanche multiplication. The maximum electric field $\bar{F}_m$ at which avalanche breakdown onsets depends on the material and on the length where multiplication takes place. The maximum field $\bar{F}_m$ is smaller in semiconductors with low bandgap energies. The electric field extends to the lower doped region in a one-sided abrupt junction. To give an example, the maximum field $\bar{F}_m$ in Si doped to $10^{17}/cm^2$ ranges to about $6*10^7$ V/m. Avalanche photodetectors (APDs) from III/V materials are widely used in high-bit-rate, long-haul optical communication systems. Compared to their p-i-n counterparts, APDs can offer better sensitivity due to their internal multiplication gain [7]. For one key figure of merit—the ionization ratio between holes and electrons—the silicon is superior to most other semiconductors. A small ionization ratio (as in Si) affects important properties like sensitivity, excess noise, and gain-bandwidth product advantageously. This is important because avalanche multiplication is a noisy process and care has to be taken that the noise floor is not increasing more than the photocurrent. Combining the good avalanche properties of Si with the absorption of different semiconductors led to heterostructure APDs.

Heterostructure APDs employ a separate absorption-charge-multiplication structure in which light is absorbed in an intrinsic heterostructure film while carriers are multiplied in the high field region of the Si film. In a Si/Ge-heterostructure APD the intrinsic absorber region is made from Ge whereas the multiplication is often made in a Si p-i-n structure. The multiplication structure consists of a thin p charge layer, a thin intrinsic multiplication layer, and an n charge layer. This high-low-high doping structure is well known in high-frequency IMPATT diodes for providing a spatially uniform avalanche multiplication rate. The full layer sequence on a Si substrate consists of an n^+-Si contact layer, an i-Si multiplication layer, a p-Si charge layer, an i-Ge absorption layer, and a p^+-top contact layer [7]. The frequency-gain product gives a characteristic performance number of APDs. In Ge-on-Si APDs, frequency-gain values of 300 GHz can be obtained with a gain of 15 and a 3 dB frequency limit of 20 GHz.

9.4 Spectral Range

The spectral range of bulk silicon stretches from the VIS to the UV. The absorption depth in UV is very short (order of magnitude 10 nm), so near-surface collection, for example, in MIS structures is necessary. This spectral range is extended to the NIR by using Ge/Si heterostructures. A further extension beyond 1.55 μm can be expected with progress in GeSn/Si heterostructures. α-Sn is a zero bandgap semiconductor. The alloy GeSn shifts the bandgap to lower values than that of Ge (0.66 eV). This material is not stable under equilibrium; sophisticated growth and device processing is necessary to exploit the NIR properties of metastable GeSn alloys.

An extension to the MIR and FIR seems not possible with bulk silicon–based materials. However, quantum size effects may be used to extend the spectral range [8–11] into the FIR. Quantum size structures like quantum wells (QWs), quantum wires, and quantum dots (QDs) break up the bands into sub-bands with quantum number $n = 1, 2 \ldots$ (Fig. 9.11).

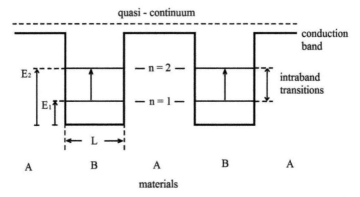

Figure 9.11 Heterostructure quantum well created by materials A/B. Shown is the conduction band (with B as lower-electron-energy material). In the quantum well electron substrates with energy edges E_1 and E_2 are created. Transitions between E_1 and E_2 are called intraband transitions.

The energies of a rectangular well are given by

$$E_n = (\hbar^2 \pi^2 / 2m^* L^2) n^2 \qquad (9.24)$$

for a well with an infinite barrier. Here m^* is effective mass, L is well width, n is quantum number, and $\hbar = h/2\pi$ is reduced Planck's constant.

To give an impression of a typical quantization energy, let's consider a well width L = 12.5 nm, and an effective mass of 0.19 m_0 (Si: transverse electron mass; m_0 free electron mass). Then we obtain E_1 = 12 meV and E_2 = 48 meV, and the intraband transition $E_1 - E_2$ would be 36 meV (corresponds to about 30 µm wavelength). So, silicon-based photodetectors may cover the whole spectral range if we include Ge/Si heterostructures and quantum size structures. The competition with other materials is especially hard in the UV (wide bandgap semiconductors), NIR (ternary III/V compounds), and FIR (small bandgap semiconductors or other device principles, like photoconduction and bolometer) range. Next we will cover the different parts of the spectrum, with a strong emphasis on NIR.

9.4.1 UV Detectors

UV light is usually divided into three wavelength regions: UV-A (320–400 nm), UV-B (280–320 nm), and UV-C (10–280 nm). UV light is harmful to human beings. It can be used in chemical and biological processes, and very hot objects (stars, rockets, etc.) and processes (welding, etc.) emit UV rays. The main applications of UV detectors are therefore in environmental, chemical, and biological analysis and in astronomy, military, and industry. The very near surface absorption is best handled with MIS photodetectors. The thin metal gate may be used as a filter. A VIS-blind design for the detector is desired because usually VIS light is hiding weaker UV emissions. An interesting demonstration was given [12] with silver gate metals as a filter having a window of around 320 nm. Ag films of thickness 70–130 nm block VIS light, whereas responsivities around 10 mA/W could be obtained at 320 nm.

9.4.2 Visible Spectrum

The VIS spectrum is the most competitive part of bulk silicon detectors. Charged coupled device–based or more recently CMOS–

based detector arrays are virtually ubiquitous in mobiles, webcams, home entrance, and security equipment. A huge industry has emerged by joining microelectronics and optoelectronics on silicon substrates. The pixel schemes are already shown in Section 9.1. Most of these detectors use a large absorption volume, which limits speed to the 100 MHz frequency region.

With SiGe/Si or Ge/Si heterostructures the photodetectors show improved frequency behavior even in the VIS spectrum. The reason is the higher absorption, which allows smaller devices.

9.4.3 NIR Spectrum

Ge on Si photodetectors are the first choice to extend the spectral range to the telecommunication wavelengths 1.3 µm and 1.55 µm. The first high-performance Ge-on-Si photodetectors were normal incidence devices for free space or optical fiber coupling. We simplify in Fig. 9.12 those parts of the band structure in Si and Ge that are responsible for NIR absorption.

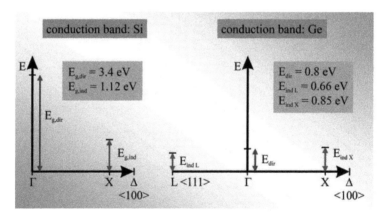

Figure 9.12 Simplified scheme of the conduction band minima in Si and Ge. Given are the energies for direct and indirect transitions. Reprinted by permission from Springer Nature Customer Service Centre GmbH: Springer Nature, *Frontiers of Optoelectronics in China*, Ref. [1], copyright (2010).

The lowest direct transition in Si is 3.4 eV (360 nm) above the valence band, whereas in Ge this direct transition comes down to 0.8 eV (1.55 µm). That means all the absorption in the VIS spectrum is indirect in Si whereas from the NIR through the VIS range, the absorption in Ge is direct. This makes Ge superior in optoelectronics in the NIR and VIS range. The lowest indirect transition in Si is the Δ minimum (Δ is near the X point, 0.8X in Si), with a gap of 1.12 eV (room temperature). In Ge the Δ minimum is at 0.85 eV, but the lowest minimum is now at the L point (111), with 0.66 eV. In unstrained Ge the Δ minimum is of small significance but in compressively strained Ge, the difference to the L minimum shrinks that in fully strained Ge on Si the Δ minimum wins (Si like conduction band structure). Note that practically fully strained Ge on Si can only be obtained for a few-nanometer thickness because of strain relaxation in highly mismatched heterosystems. A basic structure of a Ge photodiode is shown in Fig. 9.13.

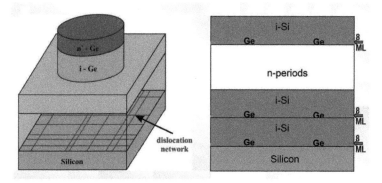

Figure 9.13 Basic structure of Ge p-i-n photodetectors on Si substrates. Left: A strain-relaxed Ge buffer layer accommodates the lattice mismatch by a dislocation network at the Si/Ge interface. Right: Ge islands arranged in planes with Si spacers (Ge-dot superlattice).

Here you see on the left a typical structure with a strain-relaxed buffer (SRB) of Ge on the Si. As a result, a dislocation network is created at the interface to the Si substrate. The combination of Si substrate and SRB is called a virtual substrate, which offers on top of a common Si substrate a different lateral lattice spacing, in this case the spacing of Ge. This virtual substrate concept is very flexible, but rather high threading dislocation densities (10^5–10^8 / cm^2) plague it. The best Ge SRBs (graded buffer) need to be about 10 μm thick, which is hardly acceptable for integrated circuits. Much more limited is the Ge amount in an approach (Fig 9.13, right) that prevents dislocations. Strain may be relaxed by surface corrugations or even better by island formation. Repeated growth of Ge island layers interrupted by Si (Ge-dot superlattice) allows dislocation-free absorber structures (as mentioned with low Ge amounts corresponding to 10–50 nm Ge layer thickness).

A sketch of a technical realization [13, 14] of a Ge photodiode is shown in Fig. 9.14.

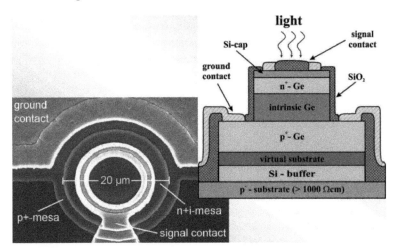

Figure 9.14 Top view (left) and cross section (right) of a Ge photodetector on a virtual substrate. © 2005 IEEE. Reprinted, with permission, from Ref. [13]. Reproduced from Ref. [14], with the permission of AIP Publishing.

On top of the virtual substrate (Si + relaxed Ge) a p-i-n structure is grown starting with a p$^+$-Ge (used as backside contact), then followed by the intrinsic Ge as the absorber layer and an n$^+$ top contact layer. For processing purposes the n$^+$-Ge top layer is covered by a thin Si cap. The top contact metal is a ring, which should be transparent in the middle for light incidence. A double-etched structure allows backside contact from the frontside and connection to high-speed CPW measurement pads. The dark current of Ge junctions is 6–7 orders of magnitude higher than in Si because of the smaller bandgap. In Ge bulk diodes current densities of 10 µA/cm^2 will be obtained. In heterostructure diodes, the dark currents [15] are usually higher because of defects (dislocations, point defects, etc.), high contact doping (Auger recombination), and thin contact layers. Figure 9.15 shows the I–V characteristics of Ge-on-Si diodes with an absorber region of different thicknesses d_i.

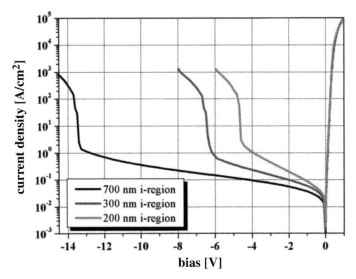

Figure 9.15 Dark current density versus bias voltage of Ge-on-Si photodiodes with different thicknesses of the intrinsic layer.

For a negative bias the reverse characteristics are shown. One sees the increase of the reverse current with a negative bias until the onset of avalanche multiplication. This onset depends on the thickness of the i-layer: −13.5 V, −6.2 V, and −4.5 V for layer thicknesses of 700 nm, 300 nm, and 200 nm, respectively. Another effect typical for defect-dominated recombination can be seen: the increase of dark current with thinner layer structures. This is typical of tunneling-assisted generation of currents from defect levels. Tunneling increases the generation as a function of the applied field. The dark currents are very similar if they are considered as a function of the field strength in the absorption layer (Fig. 9.16).

This dependence on field strength can be considered as a proof of defect level–dominated dark currents [15]. Avalanche multiplication follows the rule that breakdown field is lower in thicker layers (at lower fields the avalanche needs a longer distance for proper impact ionization).

The responsivity R (do not confuse with reflectivity) of a Ge detector is given by

$$R = R_{opt} \times \eta_{int} \times \eta_{ext}, \tag{9.25}$$

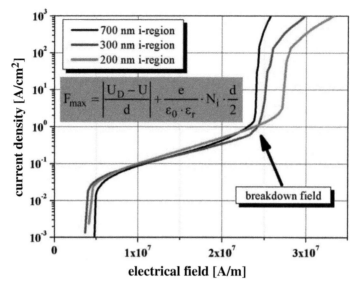

Figure 9.16 Electrical field dependence of defect-dominated dark currents in Ge-on-Si photodiodes.

with optimum responsivity R_{opt} (A/W) given by

$$R_{opt} = q/\hbar\omega = \frac{q}{hc}\lambda = \lambda(\mu m)/1.24. \quad (9.26)$$

The real responsivity is reduced by a nonideal internal quantum efficiency (η_{int} < 1) and by an external efficiency (η_{ext} < 1), which is caused by reflections and absorption in top layers [16]. A typical example of the spectral responsivity of a Ge photodiode with submicron absorption layers (thickness d_i) is shown in Fig. 9.17.

A large dark current degrades the signal-noise ratio. A figure of merit is the normalized detectivity D^* [17].

$$D^* = \sqrt{A\Delta f}/NEP, \quad (9.27)$$

where A is area, Δf is bandwidth, and NEP stands for noise equivalent power. In the case of shot noise from the dark current I_d [18] the NEP reads $(2eI_d\Delta f)^{1/2}/R$, which results in a normalized detectivity

$$D^* = \sqrt{A} \cdot R_{opt}/(2eI_d)^{1/2}. \quad (9.28)$$

The detectivity improves with responsivity and degrades with high dark currents. It is, therefore, important to operate the photodiode under low dark current conditions. As the dark current of Ge diodes is rather high, this means operation under a low reverse bias. The ideal answer would be operation under zero bias (solar cell mode).

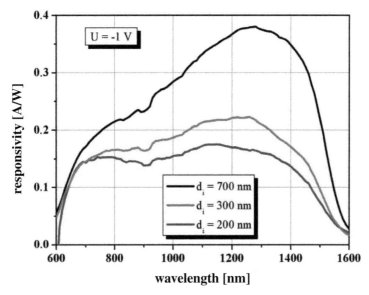

Figure 9.17 Spectral dependence of responsivity of Ge-on-Si photodiodes with different intrinsic layer widths. Republished with permission of Electrochemical Society, Inc, from Ref. [16] (1948); permission conveyed through Copyright Clearance Center, Inc.

Not only does a zero-bias operation [19] optimize detectivity, it is also beneficial for low power consumption in a stand-by modus. At this point it can be stated that the uncritical use of dark currents at a given voltage (mainly 1 V is used) is a measure of layer quality. As shown before, the dark current depends on the electrical field, which is higher in thinner layers. At the same material quality, the dark current at a fixed voltage increases with decreasing thickness. An exact measure delivers dark current versus electric field, but a rapid quality-correlated measure (number and emission properties of defect levels) would be dark current divided by thickness. For detectivity, it is crucial to find the lowest voltage for full speed operation of the photodiode.

Figure 9.18 shows the bias voltage dependence [20] of photocurrents for different optical power inputs.

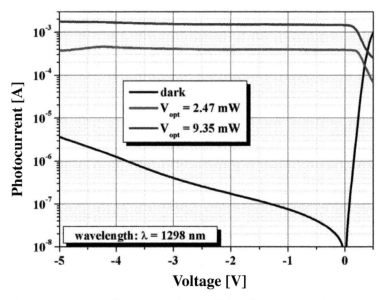

Figure 9.18 Bias voltage dependence of the photocurrent of a Ge p-i-n photodiode with 700 nm intrinsic width of the p-i-n structure. Reprinted from Ref. [20], Copyright (2009), with permission from Elsevier.

It is clearly seen that the photocurrent and, therefore, the responsivity is constant through the reverse bias ($\lambda = 1.3$ μm), at zero bias, and even within a small regime of forward bias [21]. This is a typical property of a p-i-n diode with a fully depleted i-region. Constant responsivity throughout the complete reverse bias regime does not mean that the high-frequency response stays conserved, but that is a topic of the next section in this chapter.

We have shown that for Ge p-i-n photodiodes [16] with low background doping ($<3 \cdot 10^{15}$ cm^{-3}) a zero-bias operation is possible. The responsivity curve is independent of reverse bias at $\lambda = 1.3$ μm, and this holds true for all wavelengths except the near-band-edge absorption around 1.55 μm. Here, an electro-optical effect—the Franz–Keldysh effect—modulates the absorption coefficient near the band edge with increasing field strength (Fig. 9.19).

In Fig. 9.19 $(R_{opt}/\lambda)^2$ is depicted against photon energy. In a direct semiconductor approximation for thin layers this should be a linear function near the band edge, as demonstrated in Fig. 9.19. With an increasing electrical field, the slope of the curve decreases and absorption below the direct bandgap increases. This effect is also used in absorption modulators, and details are discussed there.

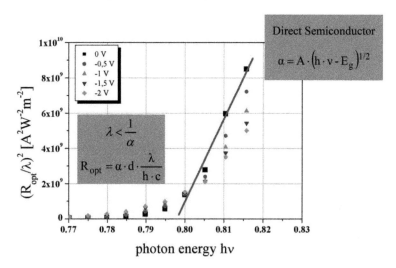

Figure 9.19 Franz–Keldysh effect in Ge-on-Si p-i-n photodiodes. Reprinted from Ref. [22], Copyright (2012), with permission from Elsevier.

We discussed the behavior of Ge-on-Si detectors for vertical incidence. The basic properties are similar for lateral waveguide incidence, with two added advantages that stem from the fact that the absorption length is now decoupled from the absorption layer thickness d_i. The length l_i of the absorption layer can be fabricated within a few micrometers to a few tens of micrometers whereas the thickness d_i stays at submicron values to facilitate monolithic integration. Two big advantages result from this decoupling: The quantum efficiency at the band edge (1.55 µm) can be increased from below 10% to nearly 100%. The design of high speed (strongly dependent on d_i) and the design of high responsivity (now dependent on l_i) can be optimized independently.

The first successful realization of a SiGe/Si-waveguide detector dates back to 1994 [23]. A bulk Si substrate with a ridge SiGe waveguide was used to couple in the wave (1.3 µm) into a pseudomorphic SiGe/Si superlattice detector (Fig. 9.20).

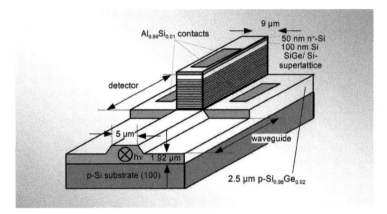

Figure 9.20 Waveguide photodetector. A rib of SiGe on Si forms the waveguide. A SiGe/Si superlattice photodiode detects the 1.3 µm light. © 1994 IEEE. Reprinted, with permission, from Ref. [23].

In the given example there is evanescent coupling from the underlying waveguide to the detector. This is possible because of the higher index of refraction of Ge leading the wave into the detector. The alternative coupling is direct butt coupling with waveguide and detector face-to-face (Fig. 9.21).

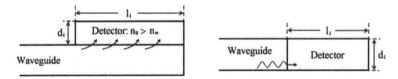

Figure 9.21 Waveguide-detector coupling. (a) Evanescent coupling mode and (b) Butt coupling.

Recent success in Ge/Si APDs with internal amplification led to a revival of this long-known device concept, which has to prove competitiveness with separate amplification with modern low noise amplifiers [6].

Ge/Si APDs have been successfully fabricated by blanket epitaxial Ge growth or selective Ge growth on epitaxial Si films [24–26]. During Ge/Si APD fabrication, a 50–100 nm p-Si charger layer can be formed either by ion implantation into the epitaxial i-Si layer or by in situ doping. Other processing techniques are similar to Ge p-i-n diodes discussed earlier.

Although the gain-bandwidth product is commonly used in literature to characterize APD performance, sensitivity more completely characterizes the overall figure of merit, which reflects the relationship between a gain-bandwidth product, responsivity, excess noise, and dark current. The theoretical sensitivity of 10 Gb/s Ge/Si APDs at a bit-error rate (BER) = 10^{-10} is a function of gain (M) for different primary dark currents (I_{DM}) [27]. Ge/Si APD receiver sensitivity is theoretically better than that of commercial III/V APD receivers (−28 to −26 dBm) [7] if optimum responsivities and a typical transimpedance amplifier noise of 1 µA are assumed.

There are two approaches to improve Ge/Si APD sensitivity: (i) reduce the primary dark current below 100 nA and (ii) enhance the primary responsivity to the commercial III/V APD level (>1.8 A/W). Similar to waveguide-coupled Ge p-i-n diodes, waveguide-integrated Ge/Si APDs offer advantages in low absolute dark current and high bandwidth-efficiency product. Recent waveguide-integrated Ge/Si APDs successfully demonstrated −31 dBm sensitivity at 1.3 µm and BER = 10^{-10} [24].

9.4.4 SiGe/Si Quantum Dot (Well) IR Photodetectors

Due to the natural valence band offset at the SiGe/Si heterojunction, discrete energy levels form inside the Si/SiGe/Si quantum confinement region. The transition between these discrete levels can be used for infrared detection [8, 11]. The SiGe/Si quantum-dot structure has been fabricated into MIS SiGe/Si QD infrared photodetectors (QDIPs) [28]. SiGe QDs were formed on a Si substrate by the Stranski–Krastanov mode [29]. It is worth mentioning that the gate electrode selected here is Al instead of Pt because for the SiGe/Si QDIP structure, the semiconductor is p-type and Pt will lead to a hole tunneling current at the inversion (positive) bias [30], which should be avoided.

The insertion of an insulator layer to fabricate the MIS photodetector indeed could reduce the dark current. For MIS SiGe/Si QDIP, the thermionic emission current of holes from Al to the semiconductor was significantly reduced due to the extra barrier and the suppression of Fermi-level pinning with an insulator layer. Boron δ doping combined with QW or QD is often used for technical realizations [31, 32].

9.5 High-Speed Operation

Usually, Si detectors are slow because they rely on the large diffusion lengths in high-quality Si. However, diffusion is a rather slow process from the random walk of carriers in the electric-field-free part of the diode. Let us now concentrate on the reason for the rapid increase of speed of Ge/Si photodetectors from the 100 MHz regime into the low millimeter-wave regime (30–50 GHz) within a few years [33–39]. The measurement by the S-parameter network analyzer includes the speed of the detector and the delay caused by the laser light modulator, so the real speed of the detector is somewhat higher, about 60–80 GHz. The essential contribution stems from the reduced device size of Ge photodiodes, which minimizes carrier transit times. An additional important measure is to suppress the slow minority carrier diffusion from absorption outside the depletion layer (Fig. 9.22).

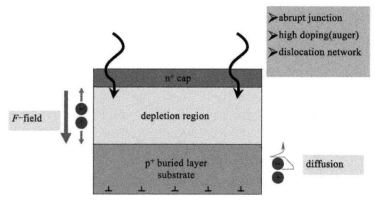

Figure 9.22 Suppression of the slow carrier diffusion from outside the depletion region.

This is accomplished by a combination of three technological steps:

1. Very abrupt junction. Transition of the highly doped contact layers to the intrinsic layer (several orders of magnitude difference in doping) is within a few nanometers. The abrupt transition confines the electric field to the intrinsic layer.
2. High doping ($>10^{20}$ cm^{-3}) of the contact layers to reduce carrier lifetime by the Auger effect. A short carrier lifetime

causes a short diffusion length, which minimizes current contributions from slow diffusion.

3. A misfit dislocation network at the bottom contact that reduces the lifetime of minority carriers. Lattice defects like dislocations push the recombination of carriers. This stops the unwanted slow diffusion of minority carriers from the bottom contact if the misfit dislocation network is outside the depletion region. (Note: Threading dislocations, which cross the depletion layer, enhance the dark current by generation of carriers in the high electric field of the depletion layer). The diffusion length L of minority carriers is given by

$$L^2 = D\tau = q\mu\tau. \qquad (9.29)$$

Here q, D, μ, and τ are the electric charge of the electron, the diffusion coefficient, the mobility, and the minority carrier lifetime, respectively.

At high doping, the diffusion length is reduced by the low mobility (a factor of 30 lower than in undoped material). The low minority carrier lifetime of high-doped semiconductors is caused by the Auger effect, where the momentum of the recombining electron (hole) is transferred to a nearby available electron (hole). As a result, the diffusion length L shrinks from more than 10 µm in a low-doped material to below 0.1 µm in high-doped layers.

Two effects dominate the detector speed if the slow diffusion is suppressed.

- The internal speed of the device is given by the transit time. Assuming a saturation velocity v_s of carriers in the depletion region, the internal frequency band f_{tr} is given by the simple expression [14]

$$f_{tr} = \frac{\sqrt{2}}{\pi} \frac{v_s}{w_D}, \qquad (9.30)$$

where v_s and w_D are the saturation velocity ($v_s = 0.6 \times 10^5/s$ in Ge) and drift width (the intrinsic width in abrupt junctions), respectively.

The assumption of saturation velocity is correct if the field strength $\bar{F}$ is more than about 3×10^6 V/m (e.g., with $V_{bi} - V = 1$ V, $w_D = 300$ nm, and the field strength $\bar{F} = 3 \times 10^6$ V/m).

- The connection to the outer world (measurement surrounding) is limited by the RC load given by the capacitance C_j of the device and by the resistance of the connection line $(R_S + 50)$ Ω (R_S is the series resistance; 50 Ω is the waveguide measurement impedance):

$$f_{RC} = \frac{1}{2\pi RC_j} = \frac{1}{2\pi} \cdot \frac{w_D}{A\varepsilon(R_S + 50)} \quad (9.31)$$

The 3 dB bandwidth f_{3dB} is roughly given by a superposition [14]:

$$\frac{1}{f_{3dB}^2} = \frac{1}{f_{tr}^2} + \frac{1}{f_{RC}^2} \quad (9.32)$$

The influence of the RC load is shown in Fig. 9.23, where a 50 GHz internal speed detector is loaded with different capacitances C_j.

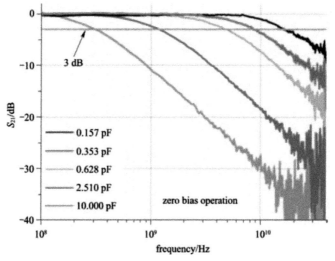

Figure 9.23 Response versus frequency for differently loaded detectors of the same internal structure (50 GHz internal speed). Reprinted by permission from Springer Nature Customer Service Centre GmbH: Springer Nature, *Frontiers of Optoelectronics in China*, Ref. [1], copyright (2010).

From these investigations, we concluded the dependence of vertical detector speed [13, 14] on the thickness of the intrinsic zone and the device mesa radius (Fig. 9.24).

The results demonstrate that with small pixel devices (4 μm diameter), speeds far above 100 GHz should be obtained readily for 100 Gbit/s detection.

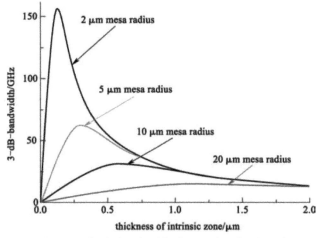

Figure 9.24 Theoretical 3 dB bandwidth of vertical Ge/Si photodetectors [14].

The speeds of properly designed Ge-on-Si detectors is already obtained at low reverse biases (Table 9.3). Ge is a small-bandgap semiconductor that is always connected with much higher dark currents than usual in Si. To minimize dark currents, operation of the device at zero bias is highly appreciated. Our group obtained very good results with a zero-bias operation [13, 33] due to the very abrupt junctions grown by molecular beam epitaxy (MBE).

Table 9.3 Three dB frequencies f_{3dB} of photodetectors (compared are detectors with reverse-bias and zero-bias detectors)

Organization	Bias/V	f_{3dB}/GHz	Wavelength/nm	Remark
IBM [39]	−4	29	850	Reverse bias
MIT [34]	−3	12.1	1540	″
LETI [35]	−2	35	1310/1550	″
USTUTT [13]	−2	39	1550	″
LUXTERA [36]	−1	> 20	1554	″
LETI [37]	−4	42	1550	″
ETRI [38]	−3	35	1550	″
USTUTT [33]	−2	49	1550	″
USTUTT [13]	0	25	1550	Zero bias
USTUTT [33]	0	39	1550	″

9.6 Outlook

Ge is a small-bandgap semiconductor with much higher dark currents than usual in Si. Recent activities with heterostructure photonic devices aim to perform competitively with photodiodes from III/V materials with the advantage of relying on sophisticated technologies from Si microelectronics. Four technological efforts have mainly contributed to the progress in Ge-on-Si photodetector performance.

These performance drivers are:

- A well-balanced trade-off of key properties by improving the responsivity at the band edge of Ge, by reducing the dark current, and by improving the linearity without compromising too much the speed.
- Extension of the wavelength window toward the MIR by using novel heterostructures that are metastable available at nanometer dimensions.
- Exploiting lateral entrance of light in waveguide geometries by carefully designed contact geometries.
- Foundry service offered for material and device technology of Ge-on-Si photonics. This allows more groups to bring in their special knowledge of photonics, and it improves the precision and reliability of processing.

The absorption of unstrained Ge decreases strongly near the direct-bandgap-cutoff wavelength of 1.55 μm. The responsivity increases, and the cutoff wavelength extends to more than 1.6 μm using tensile-strained Ge [7] with a low strain level (about 0.25%). This small strain level is simply obtained by annealing a strain-relaxed Ge-on-Si. A thermal expansion mismatch causes the tensile strain because the thermal expansion coefficient of Ge is larger than that of the Si substrate. Alternatively, a strain-relaxed GeSn buffer with small Sn amounts (1–2%) can be used if the anneal temperatures (750°C–850°C) do not fit into the process sequence. Multiple reflections enhance the absorption length, which is beneficial for vertical incidence devices. Technical solutions range from bottom-side oxide mirrors [34] to resonance cavities (resonance-cavity-enhanced photodetector) realized by dielectric Bragg reflectors on the front and bottom [40]. Light scattering on surface corrugation delivers a longer pathway in the absorber, as has already been demonstrated in solar cells. Application to photodiodes delivers a technologically simple route for responsivity enhancement [41]. The black silicon light-trapping structure can be applied to the device's rear during back end processing.

The most demanding improvement regarding detectivity concerns the dark current of Ge-on-Si photodiodes. The ideal dark current of semiconductor p/n junctions increases with small-bandgap materials, but the real dark currents of strain-relaxed structures are usually orders of magnitude higher because of defects correlated with the lattice mismatch accommodation. The dark current is mainly caused by recombination of carriers at threading dislocations and point defects correlated with dislocation climb and low growth temperatures. Carefully chosen epitaxy processes and dangling bond passivation by hydrogen have the potential to reduce the dark current densities from the order of 100 mA/cm^2 to below 1 mA/cm^2. The benchmark value for dark currents is set to 0.8 mA/cm^2 with small-bandgap III/V-epitaxy on Si substrates [42]. Space charge effects within the depletion region limit the output power of photodiodes. At high photocurrents, the linearity of the device suffers from the injection of carriers, which influences the electric field distribution. A specific doping profile in unitravelling carrier photodiodes improves the linearity to larger 1 dB saturation photocurrents [43] than available in p-i-n diodes.

Extension of the wavelength regime into the MIR is not possible with stable bulk Ge compounds. One needs either metastable GeSn alloys or highly strained Ge that is stable only in nanoscale structures. Both Sn alloying to Ge and tensile strain reduce the direct bandgap and shift, thereby, the responsivity of photodiodes beyond the 1.55 μm wavelength into the infrared region. In parallel, these material modifications reduce the energy distance between direct and indirect bandgaps. From the latter property, light emitters like LED and laser benefit whereas the first property pushes photo-detectors' wavelength range. Of utmost importance is, however, that the same technological measures support detectors and light sources. A more detailed description is given in Chapter 12. It is reasonable to expect first demonstrations of Si-based monolithic integrated photonic systems in the 1.5–2.5 μm wavelength range. The progress in material science [44] is key to getting there. In essence, two challenges face heterostructures on silicon. The most obvious is the lattice mismatch, which amounts to 4.2% for Ge on Si and to 17.4% for Sn on Ge. The lattice mismatch is below 5% for Ge-rich GeSn (Sn amount below 25%) on Ge, which is similar to SiGe/Si. Under equilibrium, high strain values are only obtained for very thin layers of a few nanometers. High strain values for thicker layers need nonequilibrium fabrication steps conserving metastable structures. Heterostructures of GeSn on Si or on Ge rely completely on metastability because the mixture of Ge and Sn breaks up into two alloy compositions under equilibrium. Fabrication at very low temperatures (below 500°C) is essential. Several approaches rival for future photonic system solutions at wavelengths beyond 2.5 µm [45]. Sharpness and defect structure of interface transitions [46] play a significant role in the optical properties of 2D (QWs), 1D (quantum wires), and 0D (QDs) heterostructures. Photodetectors from GeSn on a virtual Ge substrate are on the way toward MIR application, pending further improvement in material quality to reduce the dark current [47]. Strain-relaxed GeSn alloys with Sn amounts of 4.5% and 9.5% are needed to reduce the direct bandgap transition from 0.8 eV to 0.65 eV and 0.5 eV, respectively. The same bandgap reduction will be obtained with biaxial, tensile strain of 1% and 2%, respectively. Unfortunately, compressive strain results from the straightforward growth of GeSn-on-Ge virtual substrates, which requires high Sn amounts (>10%) for detectors with wavelengths beyond 2.5 µm. As mentioned, low-dimensional nanostructures point toward metastable GeSn alloys of good quality. Core/shell nanowires with a Ge/GeSn [48] heterostructure proved extension of the wavelength regime to 3.5 µm. The Sn content in the shell was 18%, and the compressive strain was –1.3%. The speed of vertical incidence GeSn photodetectors [49] with small areas turned out to be similar (>40 GHz) to Ge photodiodes despite the higher dark currents. High-speed operation has to be confirmed for photodiodes with higher Sn content (>4%) or high tensile strain (>0.25%).

Si-based photonics benefits strongly from the mature technology and precision of fine structure lithography. This allows rapid progress in lateral incidence waveguide photodetectors because of fabrication of small-area photodetectors with high responsivity and high speed. Lateral waveguide photodetectors decouple [7] the influence of layer thickness on responsivity and speed in contrast to vertical photodetectors, in which the responsivity increases with thickness at the cost of speed (Eq. 9.32). In lateral photodetectors, the length of the detector determines the responsivity whereas the thickness and the capacitive load (Fig. 9.23) determine the speed. Precise lithography and sophisticated contact geometries are essential to reduce the parasitic RC limitation of the speed.

Advanced CMOS concepts foresee heterostructure channel materials on Si with a focus on similar material combinations rather than on Si-based photonics, for example, Ge on Si, III/V on Si, and GeSn on Si. To tackle the challenges with such heterostructures, many instances of cooperation between industries and academia emerged and existing foundry services adopted some of these materials in their product portfolios, with astonishing performance. Some of them specialize [50] in passive and active photonic devices for waveguiding in Si and SiN and for detecting in Ge on Si.

Acknowledgments

We thank Michael Oehme for many fruitful discussions.

References

1. E. Kasper (2010). Prospects and challenges of silicon/germanium on-chip optoelectronics, *Front. Optoelectron. China*, **3**, 143–152.
2. C. Penn, T. Fromherz and G. Bauer (2000). Energy gaps and band structure of SiGe and their temperature dependence, EMIS Data Review, Vol. 24 (INSPEC, IEE, England), pp. 125–134.
3. J. Humlicek (2000). Optical functions of the relaxed SiGe alloy, EMIS Data review, Vol. 24 (INSPEC, IEE, England), pp. 249–259.
4. S. M. Sze (1981). *Physics of Semiconductor Devices*, 2nd ed. (John Wiley).
5. S. M. Sze and K. K. Ng (2007). *Physics of Semiconductor Devices*, 3rd ed. (Wiley-Interscience).
6. C. H. Lin and C. W. Liu (2010). Metal-insulator-semiconductor photodetectors, *Sensors*, **10**, 8797–8826
7. J. Michel, J. Liu and L. C. Kimerling (2010). High performance Ge-on-Si photodetectors, *Nat. Photonics*, **4**, 527–534.
8. D. Bougeard, K. Brunner and G. Abstreiter (2003). Intraband photoresponse of SiGe quantum dot/quantum well multilayers, *Physica E*, **16**, 609–613.

9. R. P. G. Karunasiri, J. S. Park and K. L. Wang (1992). Normal incidence infrared detector using intervalence-subband transitions in SiGe/Si quantum wells, *Appl. Phys. Lett.*, **91**, 2234–2236.

10. K. L. Wang, D. Cha, J. Liu and C. Chen (2007). Ge/Si self-assembled quantum dots and their optoelectronic device applications, *Proc. IEEE*, **95**, 1866–1883.

11. M. A. Gadir, P. Harrison and R. A. Soref (2001). Arguments for p-type quantum-well SiGe/Si photodetectors for the far and very-far (terahertz)-infrared, *Superlatt. Microstruct.*, **30**, 135–143.

12. W. S. Ho, C. H. Lin, T. H. Cheng, W. W. Hsu, Y. Y. Chen, P. S. Kuo and C. W. Liu (2009). Narrow-band metal-oxide-semiconductor photodetector, *Appl. Phys. Lett.*, **94**, 061114-1–061114-3.

13. M. Jutzi, M. Berroth, G. Wöhl, M. Oehme and E. Kasper (2005). Ge-on-Si vertical incidence photodiodes with 39-GHz bandwidth, *IEEE Photonics Technol. Lett.*, **17**, 1510–1512.

14. M. Oehme, J. Werner, E. Kasper, M. Jutzi and M. Berroth (2006). High bandwidth Ge p-i-n photodetector integrated on Si, *Appl. Phys. Lett.*, **89**, 071117.

15. E. Kasper and W. Zhang (2015). Device operation as crystal quality probe, in *Silicon, Germanium and their Alloys: Growth, Defects, Impurities and Nanocrystals*, G. Kissinger and S. Pizzini, eds. (Taylor and Francis, Boca Raton), pp. 341–364.

16. M. Oehme, M. Kaschel, J. Werner, O. Kirfel, M. Schmid, B. Bahouchi, E. Kasper and J. Schulze (2010). Germanium on silicon photodetectors with broad spectral range, *J. Electrochem. Soc.*, **157**, H144–H148.

17. G. H. Rieke (1994). *Detection of Light: From the Ultraviolett to the Submillimeter* (Cambridge Univ. Press).

18. C. H. Lin, C. Y. Yu, C. Y. Peng, W. S. Ho and C. W. Liu (2007). Broadband SiGe/Si quantum dot infrared photodetectors, *J. Appl. Phys.*, **101**, 033117-1–033117-4.

19. M. Oehme, J. Werner, O. Kirfel and E. Kasper (2008). MBE growth of SiGe with high Ge content for optical applications, *Appl. Surf. Sci.*, **254**, 6238–6241.

20. M. Kaschel, M. Oehme, O. Kirfel and E. Kasper (2009). Spectral responsivity of fast Ge photodetectors on SOI, *Solid-State Electron.*, **53**, 909–911.

21. M. Oehme, J. Werner, E. Kasper, S. Klinger and M. Berroth (2007). Photocurrent analysis of a fast Ge p-i-n detector on Si, *Appl. Phys. Lett.*, **91**, 051108.

22. M. Schmid, M. Kaschel, M. Gollhofer, M. Oehme, J. Werner, E. Kasper and J. Schulze (2012). Franz-Keldysh effect of germanium-on-silicon p-i-n diodes within a wide temperature range, *Thin Solid Films*, **525**, 110–114.

23. A. Splett, T. Zinke, K. Petermann, E. Kasper, H. Kibbel, H. J. Herzog and H. Presting (1994). Integration of waveguide and photodetectors in SiGe for 1.3 μm operation, *IEEE Photonics Technol. Lett.*, **6**, 59–61.

24. Y. Kang, et al. (2009). Monolithic germanium/silicon avalanche photodiodes with 340 GHz gain-bandwidth product, *Nat. Photonics*, **3**, 59–63.

25. X. Wang, et al. (2009). 80 GHz bandwidth-gain-product Ge/Si avalanche photodetector by selective Ge growth, *OSA Techn. Dig. Optical Fiber Conference*, paper OMR3.

26. N. Duan, T.-Y. Liow, A. E.-J. Lim, L. Ding and G. Q. Lo (2012). 310 GHz gain-bandwidth product Ge/Si avalanche photodetector for 1550 nm light detection, *Opt. Express*, **20**, 11031.

27. H. Kressel (1982). *Semiconductor Devices for Optical Communication* (Springer-Verlag).

28. B. C. Hsu, C. H. Lin, P. S. Kuo, S. T. Chang, P. S. Chen, C. W. Liu, J. H. Lu and C. H. Kuan (2004). Novel MIS Ge-Si quantum-dot infrared photodetectors, *IEEE Electron Device Lett.*, **25**, 544–546.

29. J. L. Liu, W. G. Wu, A. Balandin, G. L. Jin and K. L. Wang (1992). Intersubband absorption in boron-doped multiple Ge quantum dots, *Appl. Phys. Lett.*, **74**, 185–187.

30. P. S. Kuo, C. H. Lin, C. Y. Peng, Y. C. Fu and C. W. Liu (2007). Transport mechanism of SiGe dot MOS tunneling diodes, *IEEE Electron Device Lett.*, **28**, 596–598.

31. M. H. Liao, C. H. Lee, T. A. Hung and C. W. Liu (2007). The intermixing and strain effects on electroluminescence of SiGe dots, *J. Appl. Phys.*, **102**, 053520-1-353520-5.

32. C. H. Lin and C. W. Liu (2010). Metal-oxide-semiconductor SiGe/Si quantum dot infrared photodetectors with delta doping in different positions, *Thin Solid Films*, **518**, 237–240.

33. S. Klinger, M. Berroth, M. Kaschel, M. Oehme and E. Kasper (2009). Ge on Si p-i-n photodiodes with a 3-dB bandwidth of 49GHz, *IEEE Photonics Technol. Lett.*, **21**, 920–922.

34. O. I. Dosunmu, D. D. Cannon, M. K. Emsley, L. C. Kimerling and M. S. Unlu (2005). High-speed resonant cavity enhance Ge photodetectors on reflecting Si substrates for 1550-nm operation, *IEEE Photonics Technol. Lett.*, **17**, 175–177.

35. M. Rouvière, L. Vivien, X. Le Roux, J. Mangeney, P. Crozat, C. Hoarau, E. Cassan, D. Pascal, S. Laval, J. M. Fédéli, J. F. Damlencourt, J. M. Hartmann and S. Kolev (2005). Ultrahigh speed germanium-on-silicon-on-insulator photodetectors for 1.31 and 1.55 μm operation, *Appl. Phys. Lett.*, **87**, 231109.

36. G. Masini, G. Capellini, J. Witzens and C. Gunn (2008). A 1550 nm, 10 Gbps monolithic optical receiver in 130 nm CMOS with integrated Ge waveguide photodetector, *Proc. of the 5th IEEE International Conference on Group IV Photonics*, 28–30.

37. L. Vivien, D. Marris-Morini, J. Mangeney, P. Crozat, E. Cassan, S. Laval, J. M. Fédéli, J. F. Damlencourt and Y. Lecunff (2008). 42 GHz waveguide germanium-on-silicon vertical PIN photodetector, *Proc. of the 5th IEEE International Conference on Group IV Photonics*, 185–187.

38. D. Suh, S. Kim, J. Joo, G. Kim and I. G. Kim (2008). 35 GHz Ge p-i-n photodetectors implemented using RPCVD, *Proc. of the 5th IEEE International Conference on Group IV Photonics*, 191–193.

39. G. Dehlinger, S. J. Koester, J. D. Schaub, J. O. Chu, Q. C. Ouyang and A. Grill (2004). High-speed germanium-on-SOI lateral PIN photodiodes, *IEEE Photonics Technol. Lett.*, **16**, 2547–2549.

40. J. Yu, E. Kasper and M. Oehme (2006). 1.55-um resonant cavity enhanced photodiode based on MBE grown Ge quantum dots, *Thin Solid Films*, **508**, 396–398.

41. M. Steglich, et al. (2015). Ge-on-Si photodiode with black silicon boosted responsivity, *Appl. Phys. Lett.*, **107**, 051103.

42. K. Sun, et al. (2018). Low dark current III–V on silicon photodiodes by heteroepitaxy, *Opt. Express*, **26**, 13605.

43. M. Piels and J. E. Bowers (2012). Si/Ge uni-traveling carrier photodetector, *Opt. Express*, **20**, 7488.

44. E. Kasper (2016). Group IV heteroepitaxy on silicon for photonics, *J. Mater. Res.*, **31**, 3639–3648.

45. H. Lin, Z. Luo, T. Gu, L. C. Kimerling, K. Wada, A. Agarwal and J. Hu (2017). Mid-infrared integrated photonics on silicon: a perspective, *Nanophotonics*, **7**, 393–420.

46. D. J. Lockwood, X. Wu, J.-M. Baribeau, S. A. Mala, X. Wang and L. Tsybeskov (2016). Si/SiGe heterointerfaces in one-, two-, and three-dimensional nanostructures: their effect on SiGe light emission, *ECS Trans.*, **75**, 77–96.

47. X. Wang and J. Liu (2018). Emerging technologies in Si active photonics, *J. Semicond.*, **39**, 061001-1–061001-29.

48. S. Assali, et al. (2017). Growth and optical properties of direct band gap Ge/Ge0.87Sn0.13 core/shell nanowire arrays, *Nano Lett.*, **17**, 1538–1544.

49. M. Oehme, K. Kostecki, K. Ye, et al. (2014). GeSn-on-Si normal incidence photodetectors with bandwidths more than 40 GHz, *Opt. Express*, **22**, 839.

50. http://www.aimphotonics.com/pdk/

Chapter 10

Modulators

Waveguide modulator design and fabrication gives an excellent impression of how rapidly silicon-based photonic components have evolved systematically within the last decades. Very compact and highly transparent waveguides are the base of waveguide modulators. Firstly, waveguides based on silicon nitride/silicon oxide allowed miniaturization up to tenths of microns. Later, silicon/silicon oxide waveguides could be miniaturized to the submicron regime due to their high refractive index contrast. Both waveguide systems have their own merits; silicon waveguides are smaller and more densely packed, but a back-end process makes it easier to fabricate nitride/oxide waveguides. In semiconductor processing, the metallization and insulator layers above the silicon are deposited and structured under more relaxed cleanliness conditions (back end of the line [BEOL]) whereas the processing of the silicon itself requires extremely low contamination levels (front end of the line [FEOL]). Both waveguide systems are transparent in the near-infrared region but offer only a small external influence on the modulation of the light intensity. In the first generation of modulators, this drawback was overcome by a trick composed of three elements. First, the refractive index of one arm of waveguides was modulated by heating (thermo-optic effect)

Silicon-Based Photonics
Erich Kasper and Jinzhong Yu
Copyright © 2020 Jenny Stanford Publishing Pte. Ltd.
ISBN 978-981-4303-24-8 (Hardcover), 978-981-4303-25-5 (eBook)
www.jennystanford.com

or by carrier change from diode operation (plasma dispersion effect). The phase of the light wave changes at the end of an arm with a fixed length due to the modulation of the optical length (optical length is geometrical length times refractive index). In a third step, the light from this modulated arm was brought into interference by a Mach–Zehnder interferometer (MZI), with the light from a reference arm translating the phase modulation into intensity modulation. The length of the MZI modulator was rather large (centimeter range) because of the weak matter-light interaction in the transparency regime. The second generation of modulators reduced the lengths dramatically by resonant interference from coupling of light from the waveguide to ring resonators or disc resonators. The third generation of modulators utilizes a direct intensity modulation from electroabsorption effects like the Franz–Keldysh effect (FKE) in bulk semiconductors or the quantum-confined Stark effect (QCSE) in quantum wells (QWs). These electroabsorption modulators (EAMs) operate around the direct bandgap energy reasonably well. The modulation wavelength changes according to the direct bandgap with alloy composition, strain, and QW thickness. Overall, waveguide-integrated Ge and GeSi EAMs [1] have achieved a 3 dB bandwidth greater than 50 GHz, a low energy consumption of ~10 fJ/bit, and a reasonably broad (15–30 nm) operation wavelength range for multichannel wavelength division multiplexing (WDM) applications. The cause of electroabsorption is explained in Fig. 10.1 using the example of a QW. On the left side, one sees the direct-energy gap diagram of a QW with GeSi as the barrier material and Ge as the well material. The carrier wave function and the corresponding lowest energy level are marked for the valence band and the conduction band, respectively.

The bandgap of the QW increases by the quantization energy compared to the bandgap of the bulk well material. With an applied external electric field $\vec{F}$, the QW bandgap decreases but the maxima of the wave functions are spatially separated. The lower bandgap results in a redshift of the absorption, and the spatial mismatch causes a weaker increase in the absorption at higher energies. The electroabsorption also causes a redshift in the indirect band, but this is seldom utilized in devices because of the much lower absorption levels of indirect transitions. In indirect semiconductors with small energy differences between direct and indirect bandgap, like Ge and GeSn, the electroabsorption from the direct bandgap dominates.

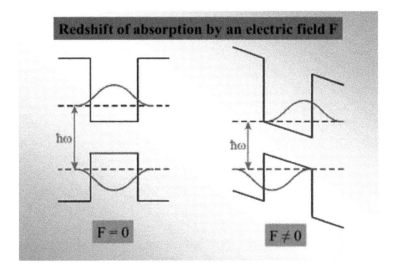

Figure 10.1 Quantum-confined Stark effect.

The modulator is the core device applied in optical cross-connect (OXC) and optical add-drop multiplexing systems; it has great application prospects in the domains of computer and optical communication. Light can be modulated on the basis of the electro-optic effect, the thermo-optic effect, the plasma dispersion effect, FKE, and QCSE, and the modulated light can carry a variety of information, especially information with huge data and high data transmission. In the following parts of the chapter, the main modulation mechanisms in group IV semiconductors and the main electrical and optical structures of silicon modulators are discussed.

10.1 Optical Modulation

There are many parameters to characterize the property of an optical field, such as amplitude (intensity), frequency, phase, and polarization. If some physical methods are used to change one of these parameters regularly with the modulation signal, the light waves can "carry" the information. This is the meaning of optical modulation. The modulation signals often refer to some applied signals, typically electrical, thermal, and optical signals. For the waveguide devices, the most common modulation forms are phase modulation and intensity modulation. Phase modulation is usually derived from a change in the refractive index of the material involved with the applied signal to change the propagation constant of light in the waveguide. However, the phase change is difficult to detect directly, which is not conducive to the extraction of modulation signals. Therefore, the most commonly used modulation form is intensity modulation, which can be achieved by changing the absorption coefficient of the material involved or converting a phase change based on phase modulation to an intensity change via interferometer or resonator structures. The latter is the often-used way for silicon-based optical waveguide modulators, which can use an MZI or a microring/microdisc resonator to convert phase modulation into intensity modulation.

The parameters of an optical waveguide modulator include modulation depth (MD), extinction ratio (ER), switching time, bandwidth, and insertion loss (IL).

Modulation depth: With the applied signal, P_{max} and P_{min} are the maximum and minimum output power of the modulator. MD is defined as the ratio between the power contrast and the maximum output power:

$$MD = (P_{max} - P_{min})/P_{max} \qquad (10.1)$$

A larger MD is better; a waveguide with an MD of 100% has the best modulation.

Extinction ratio: The ER corresponds to the MD, and it is a more commonly used parameter that is defined as:

$$ER = 10 \log(P_{max}/P_{min}) \qquad (10.2)$$

Switching time: The switching time is an important parameter for modulator performance. Imposing a square-wave signal in the modulator, the output is also a square-wave signal. The difference between the maximum and minimum output power is defined as $\Delta P = P_{max} - P_{min}$. The rise time t_r is the time experienced by the output power to increase from $P_{min} + 10\%\Delta P$ to $P_{min} + 90\%\Delta P$. Similarly, the fall time t_f is the time experienced by the output power to decrease from $P_{max} - 10\%\Delta P$ to $P_{max} - 90\%\Delta P$. Hence, the switching time t_s of a modulator is the larger value of rise time or fall time.

Modulation bandwidth: Modulation bandwidth can be defined as the modulation frequency while the ER reduces to half of the maximum. It is often associated to the switching time t_s:

$$f_{3dB} = 1/t_s \qquad (10.3)$$

Insertion loss: It is the total loss of the entire modulator, including the coupling loss, the propagation loss in the waveguide, and the absorption loss in the modulation region. If the input power and output power are P_{in} and P_{out}, respectively, its IL can be written as:

$$IL = 10\log[(P_{in} - P_{out})/P_{in}] \qquad (10.4)$$

10.2 Modulation Mechanisms

Let us first introduce the modulation mechanisms by changing the effective refractive index in silicon. There are many physical interactions to modulate light in silicon, such as thermo-optical, electro-optical, and optical-optical (which includes linear and nonlinear optics) effects. A number of mechanisms, such as the electro-optic effect, the thermally induced index change, the carrier-injection-induced index change (carrier plasma effect), and electroabsorption, classify the interaction. In this section, we will first introduce the modulation mechanisms in group-IV semiconductors.

10.2.1 The Electro-optic Effect

Many optical materials demonstrate an optical anisotropic effect with the applied electric field, including the Pockels effect and the Kerr effect. The electro-optic effect refers to the change in the refractive index of the material with the applied electric field $\vec{E}$. The refractive index change caused by the applied electric field can be expressed as

$$n = n_0 + a\vec{E} + b\vec{E}^2 + \ldots, \qquad (10.5)$$

where n_0 is the refractive index of the material without the applied electric field. It is the Pockels effect that the refractive index linearly changes with the electric field, and a is the linear electro-optic coefficient. It is the Kerr effect that the refractive index second-orderly changes with the electric field, and b is the quadratic electro-optic coefficient. The Pockels effect exists only in the noncentrosymmetrical crystal, such as gallium arsenide (GaAs) and lithium niobate (LiNbO$_3$); their Pockels coefficients are 1.6×10^{-10} cm/V and 3.08×10^{-9} cm/V, respectively. These electro-optic crystals are ideal materials for modulators and can easily be used to achieve low-power and high-speed modulators in discrete arrangements.

Monocrystalline silicon is a typical centrosymmetrical crystal with no Pockels effect and a weak Kerr effect [2]. The refractive index in silicon has a change of 10^{-5} with an applied electric field of up to $3*10^5$ V/cm. This change has been able to meet the requirement of the modulator, but the voltage is near the breakdown voltage of lightly doped silicon, which limits the use in low-power applications.

Another electro-optic effect is connected with the Franz–Keldysh effect (FKE) [3, 4]. With the applied electric field, the energy band in silicon tilts so that electron and hole wave functions extend into the bandgap. The transitions of below-the-bandgap energy may occur with the influence of the photon-associated tunneling as shown in Fig. 10.2a. If the energy of the incident light is close to the bandgap, the absorption spectrum will shift to the long wavelength with the applied electric field, which can be used to fabricate electroabsorption modulators (EAMs) by direct intensity modulation.

A change in the refractive index following from a change of absorption coefficient (Kramers–Kroning rule) will dominate indirect bandgaps, which can be used for phase modulation. Figure 10.2b shows the changes in refractive index connected with the FKE. It has been demonstrated that the change in the refractive index caused by the FKE is about 1.3×10^{-5} when wavelength and applied electric field are 1.07 μm and 10^5 V/cm.

Note that the influence of the FKE both for absorption and for refractive index is strong only around the bandgap energy [5]. The indirect bandgap energy of silicon is about 1.12 eV at room temperature. Optical transitions from an indirect bandgap need a momentum transfer from photons, which results in two cut-off energies—1.06 eV and 1.18 eV—for the optical phonon replica (optical phonon energy in Si: about 60 meV).

The band structure of SiGe alloys up to high Ge amounts (80%) is similar to that of Si; only the absolute value of the bandgap energy decreases from 1.12 eV to 0.83 eV. Accordingly, phase modulation based on refractive index changes from FKE may be used with SiGe alloys between about 1.1 eV and 0.8 eV.

The direct bandgap transition of Ge and Ge-rich SiGe alloys dominates the optical absorption behavior of these indirect materials with a small energy difference to the direct transition. For these materials the direct intensity modulation utilizing the absorption redshift [6] of the FKE is more attractive.

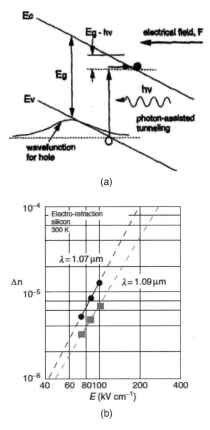

Figure 10.2 (a) the Franz–Keldysh effect causes a tilting energy band in silicon with the applied electric field; (b) the relationship between the changes in the refractive index of silicon and the applied electric field with wavelengths at 1.07 μm and 1.09 μm. © 1987 IEEE. Reprinted, with permission, from Ref. [2].

Alloy layers and quantum wells (QWs) from GeSn extend the wavelength regime of FKE modulators beyond the 1.55 μm band [7].

10.2.2 The Thermo-optic Effect

The material refractive index n changes with the temperature T, which is the thermo-optic effect. The temperature increases lead

232 | Modulators

to an increase in the effective index in silicon. Silicon has a large thermo-optical coefficient ($\partial n/\partial T \approx 1.86 \times 10^{-4}$ K^{-1}) of 1.55 µm, which is about 2 times and 15 times larger than that of LiNbO$_3$ and SiO$_2$, respectively. Silicon also has large thermal conductivity ($\sim 1.4 \times 10^5$ W/K). We can fabricate modulators with simple structures by using the thermo-optic effect. However, the response speed of the thermo-optic effect is relatively slow; for the submicron SOI optical modulator, the fastest switching time can just reach microsecond timescales. The thermo-optic effect is very useful for principal investigation of modulators and for modulator fine adjustment of operation at a certain wavelength.

10.2.3 The Plasma Dispersion Effect

The plasma dispersion effect means that the changes in the concentration of free carriers lead to the changes in the refractive indexes and absorption coefficients of semiconductor materials. Silicon has a notable plasma dispersion effect (carrier injection or depletion). It has been demonstrated that we can achieve a high speed modulation in this effect. The effect currently is the main operating foundation for silicon optical modulator.

We can derive the expressions for the refractive index and absorption coefficient in silicon as follows, assuming an effective mass model for the bandgap:

$$n = n_0 - \frac{e^2 \lambda_0^2}{8\pi^2 \varepsilon_0 n_0 c^2} \left[\frac{N_e}{m_e^*} + \frac{N_h}{m_h^*} \right] \tag{10.6}$$

$$\alpha = \frac{e^3 \lambda_0^2}{4\pi^2 \varepsilon_0 n_0 c^3} \left[\frac{N_e}{(m_e^*)^2 \mu_e} + \frac{N_h}{(m_h^*)^2 \mu_h} \right] \tag{10.7}$$

Here n and n_0 are the refractive indexes with and without carrier injection, respectively; N is the concentration of the free carriers, m^* is the effective mass, μ is the mobility, c is the velocity of light, and the subscripts e and h represent electron and hole, respectively. The above two equations are the corresponding relation between the refractive index and absorption coefficient in the semiconductor material, namely the free carrier plasma dispersion effect of the material. Going by Eqs. 10.6 and 10.7, if the concentration of the free carrier increases, the refractive index will decrease but the absorption will be enhanced. The refractive index and the absorption coefficient can be simultaneously changed through a change in the concentration of the free carrier. Therefore, we can make use of the changes in the refractive index (Fig. 10.3) and absorption (Fig. 10.4) to change the phase and intensity of light, achieving light modulation.

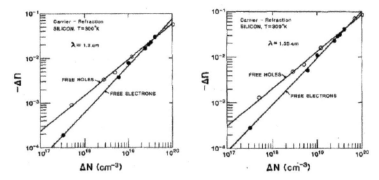

Figure 10.3 Refractive index change with the concentration of free carriers. (a) λ_0 = 1.3 μm and (b) λ_0 = 1.55 μm. © 1987 IEEE. Reprinted, with permission, from Ref. [2].

Figures 10.3 and 10.4 show the variation in the refractive index and absorption coefficient in silicon with a change in the concentration of the free carrier at the wavelengths of 1.3 μm and 1.55 μm. Soref and Bennett [2] used the Kramers–Kroning dispersion relation and experimental data for the absorption spectrum to obtain approximate expressions of the free carrier plasma effect in silicon. The experimental values (Eqs. 10.8 and 10.9) deviate in detail from the simple bandgap model (Eqs. 10.6 and 10.7) because of the real sub-band structure with different effective masses.

$$\lambda_0 = 1.3 \, \mu m \Rightarrow \begin{cases} \Delta n = -\left[6.2 \times 10^{-22} \Delta N_e + 6.0 \times 10^{-18} (\Delta N_h)^{0.8} \right] \\ \Delta \alpha = 6.0 \times 10^{-18} \Delta N_e + 4.0 \times 10^{-18} \Delta N_h \end{cases}$$

(10.8)

$$\lambda_0 = 1.55 \, \mu m \Rightarrow \begin{cases} \Delta n = -\left[8.8 \times 10^{-22} \Delta N_e + 8.5 \times 10^{-18} (\Delta N_h)^{0.8} \right] \\ \Delta \alpha = 8.5 \times 10^{-18} \Delta N_e + 6.0 \times 10^{-18} \Delta N_h \end{cases}$$

(10.9)

Here Δn is the change in the refractive index, $\Delta \alpha$ (cm^{-1}) is the change in the absorption coefficient, ΔN (cm^{-3}) is the change in free carrier concentration, and the subscripts e and h represent electron and hole, respectively. Comparing with the Kerr effect and the FKE, the changes in the refractive index caused by the carrier plasma effect are much larger for indirect bandgap semiconductors. It was the primary choice to achieve silicon electro-optic modulator.

Direct intensity modulation from the electroabsorption effect enlarged the modulator toolbox since heterostructures on Si are available in research and foundry service.

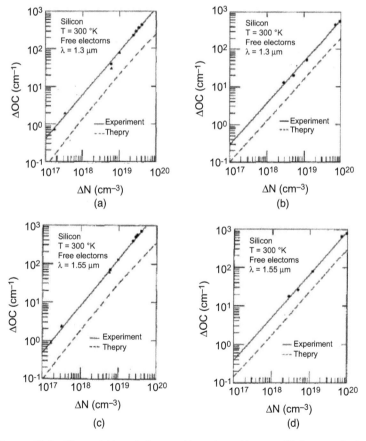

Figure 10.4 The relation between the absorption coefficient and the concentration of the free carrier (electron and hole) in theory and experiment. (a) Electrons, 1.3 μm. (b) Holes, 1.3 μm. (c) Electrons, 1.55 μm. (d) Holes, 1.55 μm. © 1987 IEEE. Reprinted, with permission, from Ref. [2].

10.2.4 Strain and Quantization Effects

10.2.4.1 Strain-induced linear electro-optic effect

Monocrystalline silicon has no linear electro-optic effect because it is a centrosymmetrical crystal. However, a significant linear electro-optic effect can be induced in silicon by breaking the crystal symmetry. The symmetry is broken by depositing a straining layer on top of a silicon waveguide [8]. When a straining layer of SiO_2 or Si_3N_4 is deposited on top of an SOI waveguide, it will expand the structure underneath in both directions horizontally. So the crystal symmetry is broken, which leads to a linear electro-optic coefficient in silicon, and the refractive index will have linear changes with the applied electric field. The linear electro-optic coefficient induced by strain in silicon is measured to be 15 pm/V. It is much smaller than the linear electro-optic coefficients induced by the commonly applied electro-optic crystals, such as $LiNO_3$ (360pm/V), but is comparable with those induced by other noncentrosymmetric semiconductors, like GaAs. In addition, the waveguide structure with a large group refractive index can increase the electro-optic coefficient of strained silicon. Photonic crystal waveguide often has a large group refractive index. It has been demonstrated that silicon can obtain the electro-optic coefficient of ~830 pm/V in the photonic crystal waveguide.

High strain values are obtained by lattice-mismatched heterostructures. Strain shifts considerably the bandgap energies [9], which allows an adjustment of operating wavelengths.

10.2.4.2 Quantum-confined Stark effect

The quantum-confined Stark effect (QSCE) is exhibited when an applied electric field is perpendicularly added on the top of QW structures [10]. It is the quantum-confined version of the FKE in bulk materials. The intensity is larger because the thin QW forces the wave function overlap between hole and electron states even under strong electric fields. Thin QW structures made from direct bandgap III–V semiconductors such as GaAs and InP and their alloys exhibit a strong Stark effect [11] mechanism, which allows modulator structures with only several microns of optical path length. The QCSE, at room temperature, in thin germanium QW structures grown on silicon has been reported [12], which causes the absorption spectrum of the QW to red-shift to a long wavelength with the applied electric field. The QCSE is widely used in the III–V QW structures (such as InGaAs/InAlAs) to fabricate the EAM. However, the QCSE is usually very weak in silicon because of the indirect bandgap. However, a very strong QCSE of the direct transition exists in the GeSi/Ge QW structure. The maximum effective absorption coefficient change in this QW structure is 2800 cm^{-1} at 1438 nm under a 3 V bias. Going by the result, the structure is very promising for small, high-speed, low-power EAMs. Low energy requirement in datacom pushes the use of EAM based on the FKE and the QCSE. All this success [13] indicates that GeSi EAMs are well on the way to system integration in small-footprint, ultra-low-energy, high-speed photonic datalinks.

10.3   Structures for the SOI Waveguide Modulator

It has been widely acknowledged that the plasma dispersion effect and the electroabsorption are the fastest modulation mechanisms for the silicon-based electro-optic modulator. The silicon electro-optic modulator based on the plasma dispersion effect is usually composed of two parts: an electrical structure and an optical structure [14–25]. To change the carrier concentration distribution, many different electrical structures can be used to achieve the injection, accumulation, depletion, and inversion of free carrier, so the refractive index (or absorption coefficient) can be changed correspondingly. There are three commonly used electrical structures: a forward-biased p-i-n diode, a reverse-biased p/n junction, and a metal-oxide-silicon (MOS) capacitor. Optical structures are mainly used for guiding the light and converting phase modulation to intensity modulation. The commonly used optical structures can be classified as interference, total internal reflection, and absorption, where interference also can be classified as the Mach–Zehnder interferometer (MZI), the Fabry–Perot (F-P) resonator, the microring resonator (MRR), and the photonic crystal waveguide.

10.3.1   Electrical Structures

10.3.1.1   Forward-biased p-i-n diode structure

Figure 10.5 shows the forward-biased p-i-n structure in the waveguide modulator: the waveguide core region is the lightly doped intrinsic region, and the top of rib region or slab region on both sides of the rib region are the heavily doped P⁺ region and N⁺ region, respectively. The forward-biased p-i-n diode is a simple bipolar electrical unit and also the most commonly used electrical structure. With a forward bias, a large number of nonequilibrium carriers are injected into the intrinsic region. Because of the carrier plasma dispersion effect, the refractive index and absorption coefficient in the rib waveguide are changed simultaneously. The nonequilibrium carriers in the high-impedance lightly doped silicon have a certain diffusion length, so there is a relative large overlap between the varying region of the refractive index and the optical field region, which causes high modulation efficiency.

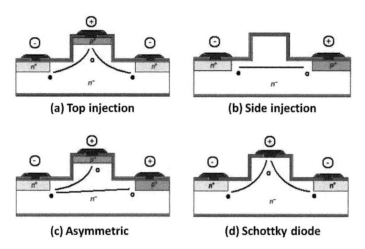

Figure 10.5 Different forward-biased p-i-n structures in the waveguide modulator.

There are four main forms for p-i-n electrical structures in the rib waveguide, as shown in Fig. 10.5, where the p- and n-doped types can be exchanged with each other. The physical factors of the carrier injection, drift, diffusion, absorption, and loss should be considered for design to enhance the modulation efficiency and speed as much as possible. The lateral-injection p-i-n diode structure is the most tolerant to fabrication technology, and it also can enhance the modulation speed by reducing the distance between heavily doped regions on both sides of the rib; hence, the lateral-injection p-i-n diode structure is used often for device design.

10.3.1.2 Reverse-biased p/n junction structure

The waveguide cross section of the modulation region of the electro-optic modulator based on a reverse-biased p/n junction is shown in Fig. 10.6. It is composed of a p/n junction and two electrode layers of p$^+$ and n$^+$ without the i-region, like in the above structure. The p/n junction operates with a reverse bias. The medium-doped p-type and n-type silicon in the waveguide core region compose a horizontal or vertical p/n junction. There is a certain carrier distribution in the waveguide without the applied modulation; however, if a reverse bias is imposed, the width of the depletion region will increase and the concentration of carriers around the p/n junction will decrease; then the refractive index in the depletion region will be changed by the changes in the concentrations of carriers.

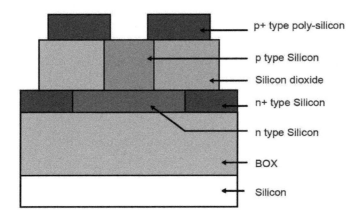

Figure 10.6 Cross-section schematic of a silicon modulator based on a reverse-biased p/n junction structure.

The p/n junction is in the state of carrier depletion with a reverse bias, and carriers mainly drift in the waveguide with the applied electric field, so it has a relative high response speed. Theoretical analysis shows that this modulator can achieve a high-speed modulation of up to several hundred gigahertz. However, its modulation efficiency is relatively low because of limited extension of the depleted region and its small overlap with the optical field. Simultaneously, absorption in the waveguide increases due to a doped core layer, so the propagation loss of the device correspondingly increases. The modulation efficiency of this structure is difficult to enhance directly, but with strong resonant structures, such as MRRs, it can reduce the amount of refractive index changes in the requirement of electro-optic modulation, which helps achieve better modulation effects. In addition, the modulation speed can be enhanced further.

10.3.1.3 MOS capacitor structure

Figure 10.7 shows the cross-sectional view of a silicon electro-optic modulator based on an MOS capacitor [20]. The first high-speed silicon modulator with a modulation bandwidth exceeding 1 GHz based on this structure was demonstrated. It comprises an n-type doped crystalline silicon slab (the silicon layer of the SOI wafer) and a p-type doped polysilicon rib with a thin gate oxide layer sandwiched between them. With the applied bias, free carriers accumulate on both sides of the gate oxide layer, and then the changes in the refractive index caused by the changes in the carrier concentration achieve the optical modulation in the device.

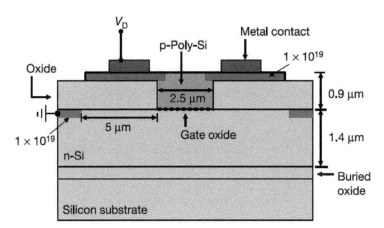

Figure 10.7 The cross section of a silicon modulator based on an MOS capacitor. Reprinted by permission from Springer Nature Customer Service Centre GmbH: Springer Nature, *Nature*, Ref. [20], copyright (2004).

The carriers drift rather than inject or diffuse in the MOS capacity structure waveguide with the applied electric field, so it can obtain a high-speed modulation in the structure. The modulation speed of the device is mainly dependent on the resistance in the waveguide layer between electrodes and the capacity of the gate oxide layer. Free carriers move in the device controlled by voltage rather than current, so the power consumption and modulation interference by the electro-optic effect are relatively small in the device. However, the region in which carriers accumulate is relative small due to the thin gate oxide layer and the corresponding overlap with optical field is also relatively small, so its total modulation efficiency is not high. The device has strong polarization dependence due to the introduction of the gate oxide layer; the modulation efficiency of TE mode is much larger than the one of TM mode. It is more important that the modulation structure has high requirements for fabrication technology. In order to reduce power consumption and enhance modulation efficiency, the gate oxide layer should be thin as much as possible when ensure it not to be broken down with the applied voltage. The interface between the gate oxide and silicon should be smooth as much as possible to reduce the propagation loss of optical field. The quality of polysilicon on the top of gate oxide is also related to the propagation loss of optical field. Therefore, although the structure can achieve a high speed modulation, it is difficult to fabricate and less tolerant to fabrication technology.

Table 10.1 shows the comparison of the three electrical structures.

Table 10.1 Comparison of the three electrical structures

Electrical structure	Forward biased p-i-n diode	Reverse biased PN junction	MOS capacitor
Modulation efficiency	High	Low	Low
Fabrication	Easy	Difficult	Difficult
Power consumption	Large	Small	Small
Influence of thermal-optic effect	Large	Small	Small
Optical absorption	Weak	Strong	Strong
Modulation speed	Low	High	High

10.3.2 Optical Structures

10.3.2.1 Mach–Zehnder interferometer structure

The MZI structure is composed of the input/output waveguides, two 3 dB couplers (a splitter and a combiner), and two modulation arms (Fig. 10.8). The incident light beam is split into two beams by the splitter. They enter the two modulation arms, propagate a certain distance in the modulation arm, and finally combine into a single beam to provide the output light. A certain optical path difference will be induced through a change in the refractive index of one of the two arms. This optical path difference modulates the magnitude of phase difference between both beams, and the magnitude of phase difference determines the optical field distribution that is generated after the two beams interfere through a combiner. Therefore, it is possible to achieve the modulation or switching function of light by changing the refractive index of one of the two arms. Light propagating in the MZI achieves a conversion from phase modulation to intensity modulation.

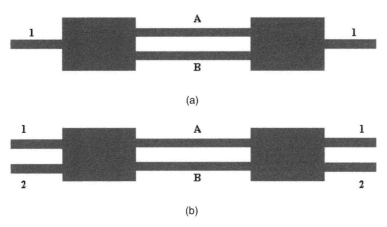

Figure 10.8 Two main MZI configurations: (a) 1 × 1 MZI; (b) 2 × 2 MZI.

The operation principle of the 1 × 1 MZI is relatively simple. Let us take 1 × 1 MZI as an example to derive the operation principle of the MZI. Assuming the splitter in the input port and the combiner in the output port are symmetrical, the splitting ratio of the splitter in the input port is $a:(1 - a)$, the additional loss factor is α, the phase difference of light in the two output ports is $\Delta\varphi$, the lengths of the two modulation arms are L_A and L_B, the effective refractive indices in the two modulation arm waveguides are n_{effA} and n_{effB}, and the propagation losses are a_A and a_B. After light of power P_{in} enters from the input port and goes through the splitter, the input optical fields in both arms can be expressed as:

$$E_{1A} = \sqrt{\alpha a P_{in}} \exp[j(\omega t + \Delta\varphi_s)] \qquad (10.10)$$

$$E_{1B} = \sqrt{\alpha(1-a)P_{in}} \exp[j\omega t] \qquad (10.11)$$

At the end of the two arms, the output optical field can be expressed as:

$$E'_{1A} = \sqrt{\alpha a P_{in}} \exp[j(\omega t + \Delta\varphi_s)]\exp(-jk_0 n_{effA} L_A - \alpha_A L_A)$$
$$\qquad (10.12)$$

$$E'_{1B} = \sqrt{\alpha(1-a)P_{in}} \exp[j\omega t]\exp(-jk_0 n_{effB} L_B - \alpha_B L_B) \qquad (10.13)$$

The output port is a 2×1 combiner, its output optical field is the interference between the two input fields E'_{1A} and E'_{1B}, and the additional loss factor β, then we can get the output optical field

$$E_{11} = \sqrt{\beta}\{\sqrt{\alpha a P_{in}} \exp[j(\omega t + \Delta\varphi_s)]\exp(-jk_0 n_{effA} L_A - \alpha_A L_A)+$$
$$\sqrt{\alpha(1-a)P_{in}} \exp[j\omega t]\exp(-jk_0 n_{effB} L_B - \alpha_B L_B)\}$$
$$\qquad (10.14)$$

Therefore, the optical power in the output waveguide is

$$P_{out} = |E_{11}|^2 = \alpha\beta P_{in}\{a\exp(-2\alpha_A L_A)+(1-a)\exp(-2\alpha_B L_B)+$$
$$2\sqrt{a(1-a)}\exp(-\alpha_A L_A - \alpha_B L_B)\cos[\Delta\varphi_s - k_0(n_{effA} L_A - n_{effB} L_B)]\}$$
$$\qquad (10.15)$$

From the above analysis, the factors of the beam splitting, the beam combining, the balance between optical fields in both arms, and the phase difference in the MZI all affect the interference and superimposition of the output optical field, and then directly determine the corresponding modulator operation characteristics.

The MZI optical structure can obtain a high ER and a wide optical bandwidth, and the structure is easy to fabricate and has a large fabrication tolerance, but the refractive index change caused by the plasma dispersion effect is often very small. A very long modulation

Structures for SOI Waveguide Modulator | 243

arm is required to achieve a π phase shift, so the length of the device based on the MZI structure may be very long, up to the millimeter order of magnitude.

10.3.2.2 Fabry–Perot resonator structure

The resonator structure usually can confine the optical field in a small region, so its propagation characteristics are very sensitive to the refractive index changes in the resonator. Based on the principle, even though the refractive index change caused by the plasma dispersion effect is very small, the silicon-based modulators also can be achieved with small dimensions. There are many common used optical resonator structures based on silicon waveguides, including F-P resonators, MRRs, microdisc resonators.

An F-P resonator is a resonator structure widely used in laser. It typically constitutes of two parallel highly reflective mirrors and a medium (or waveguide) between both mirrors, as shown in Fig. 10.9a. When an incident light travels into the left reflective mirror, a part of light is reflected and the other part of light is transmitted. The transmitted light travels through the medium (or waveguide) between both mirrors with a certain phase shift, and then travels into the right mirror; similarly, a part of light is reflected and the other part of light is transmitted. The partial reflected light travels in the medium (or waveguide) in the reverse direction again, and then is reflected by the left mirror. Again and again, there exists the superposition of multiple transmitted lights in the output port. They have a fixed phase difference that is related to light wavelength and optical cavity length. For some specific wavelengths, transmitted light in the output port is in-phase while for other wavelengths, the transmitted light in the output port is out of phase. Therefore, the transmission characteristics of the F-P resonators have resonance peaks. Figure 10.9b shows the transmission spectrum of an F-P resonator based on a silicon waveguide.

If carriers are injected into the waveguide between both reflective mirrors, the refractive index in the waveguide will change, resulting in the change in the optical cavity length of F-P resonators, so the resonance peaks in the transmission spectrum change, as shown in Fig. 10.9b. For a certain fixed wavelength, the transmissivity of total F-P cavity changes, which cause the change in the output intensity, so it can achieve the modulation function. The modulation speed of F-P modulators is related to not only electrical structures but also optical transmission characteristics. The modulation is achieved mainly by the shift of resonance peaks, if the quality factor (Q) in the F-P cavity is larger, the resonance bandwidth will be narrower, and the refractive index change, required for shifting the same wavelength, is smaller. Therefore, for the same ER, a modulator based on an F-P resonator of high Q can achieve a short switching time.

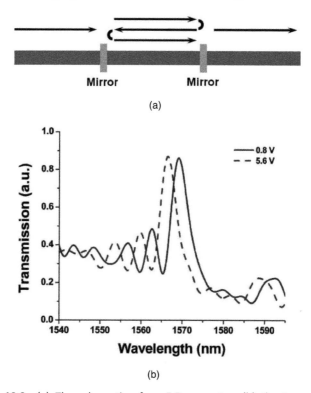

Figure 10.9 (a) The schematic of an F-P resonator; (b) the transmission spectrum of an F-P resonator based on silicon waveguide. Reprinted with permission from Ref. [14] © The Optical Society.

10.3.2.3 Microring resonator structure

Marcatili [26] first proposed the optical MRR in 1969. Recently, MRR has been attracted wide attention in silicon-based photonic devices by researchers. An MRR is an important optical structure with wavelength selectivity, so it can be used to control the light transmission path; and the optical field intensity in the resonant cavity is very high, so it also can be used for many nonlinear optics experiments. An MRR is an ideal basic building block for very-large-scale-integrated (VLSI) optical circuits, which has been widely used in fabricating a serial of passive or active optical functional devices, such as add-drop filters, multiplexers/de-multiplexers, lasers, modulators, switches, gates, delay lines, sensors, etc.

An MRR comprises a ring or racetrack waveguide and one or two straight waveguides coupled with the ring or racetrack waveguide, as shown in Fig. 10.10a. Unlike the F-P resonator, the transmitted light and reflected light export from different waveguides: the transmitted light travels through the through port and the reflected light travels through the drop port; hence, with MRRs, it is easier to achieve optical switches with multi-input and multi-output. Light can be coupled into the ring via an evanescent field through the input waveguide. The ring will be resonant for light whose phase change after each full round trip in the ring is an integer multiple of 2π. The light coupled into the ring from the input waveguide will be in phase with the original light in the ring and constructive interference, on the other hand, the light coupled back into the input bus will be reverse-phase with the original light in the input bus and destructive interference. The light in the ring can achieve a very high intensity and export through the drop port by coupling into the output bus. While the phase change after each full round trip in the ring is not an integer multiple of 2π, light coupled into the ring will be very small and most of them transmit directly through the through port, that is, the MRR is transparent to the nonresonant light.

An MRR is functionally the same as the F-P resonator, however, comparing with the F-P resonator, the quality factor Q of an MRR is much higher, so it is more conducive to achieve a high speed electro-optic modulator. Its output spectrum is similar to the F-P with resonance peaks. The resonance peaks can be shifted by changing the refractive index in the ring waveguide via electrical structures, and then the modulation in the MRR structure is achieved. Figure 10.10b shows the output spectrum in the through port of an MRR where an electrical structure of forward biased p-i-n diode is embedded to inject carriers. The solid line represents the output spectrum in the through port without the injection of carriers, and the dotted lines represent the output spectrums in the through port with different applied biases. It can be clearly seen that the output optical power in the through port increases, while the applied bias increases for some specific wavelengths.

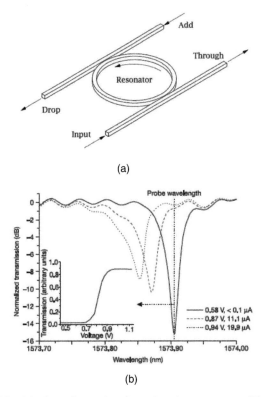

Figure 10.10 (a) The schematic of a microring structure; (b) the output spectrum of a ring resonator. Reprinted by permission from Springer Nature Customer Service Centre GmbH: Springer Nature, Nature, Ref. [17], copyright (2005).

If a disc waveguide is used to replace the ring waveguide in the MRR, the microdisc resonator is generated. A microdisc resonator can achieve a higher Q, and its optical response speed is much higher than electrical response speed, so it is expected to achieve a higher-frequency modulator.

Both the F-P resonator and the microring (microdisc) resonator can greatly reduce the size of modulators to micrometer order of magnitude, which is a great advantage for large scale optoelectronic integration. However, resonator modulators can only operate in the specific wavelength range with a narrow wavelength band, which requires a carefully consideration on the operating wavelength of device for the design. In addition, because the refractive index change is very sensitive in the resonator structure, some parasitic effects have an impact on the performance of electro-optic modulators. For instance, the device based on the electrical structure of p-i-n diode generates heat due to the injection of current. However, the thermo-optic effect and plasma dispersion effect is competing with each other, and their refractive index changes is just opposite. If the device design is not reasonable, the plasma dispersion effect may mask the electro-optic effect.

10.4　Summary

Two technical directions turned out to be dominating research and applications of modulators in Si-based photonic systems. (i) First, Si waveguide modulators based on phase modulation from the plasma dispersion effect matured utilizing different CMOS technologies [27] for resonant interference (microring, microdisc) in small-footprint devices. (ii) Heterostructures (Si/Ge and GeSn/Ge) that are necessary for detectors and light sources paved the way for recent progress in EAMs, which give direct intensity modulation from Franz–Keldish and QCSE [1, 13]. System integration of Ge/GeSi EAM has gained momentum driven by the energy efficiency requirements in datacom. Recent highlights are high speed greater than 50 GHz, low energy consumption ~10 fJ/bit, and a reasonably broad (15–30 nm) operation wavelength range for multichannel wavelength division multiplexing (WDM) applications. Research groups [28, 29] demonstrated a 16-channel GeSi EAM-photodiode array coupled to multicore fibers to achieve 896 Gbps for short-reach optical links, and they demonstrated a 100 Gbps nonreturn to zero (NRZ) on-off key (OOK) transmission format using GeSi EAM. The direct intensity modulation by EAM is especially integration friendly because the electrical and optical structures for detection, light emission, and modulation are identical. The electrical structure is given by a diode [30] that is operated for detection at zero bias or at small reverse bias. For light emission the diode is driven in the forward direction, and for modulation the diode is switched between reverse bias and zero bias. The optical structure needs coupling between the waveguide and the active device that is done by butt coupling or evanescent coupling, same for all three devices.

Acknowledgments

We thank Xuejun Xu and Kang Xiong for the collection of basic information and figure drawings.

References

1. X. Wang and J. Liu (2018). Emerging technologies in Si active photonics, *J. Semicond.*, **39**, 061001-1–061001-29.

2. R. A. Soref and B. R. Bennett (1987). Electrooptical effects in silicon, *IEEE J. Quantum Electron.*, **23**, 123–129.

3. W. Franz (1958). Einflus eines elektrischen Feldes auf eine optische Absorptionskante, *Z. Naturforsch.*, **13a**, 484–489.

4. L. V. Keldysh (1957). Behaviour of non-metallic crystals in strong electric fields, *JETP*, **33**, 994–1003.

5. A. Frova, P. Handler, F. A. Germano and D. E. Aspnes (1966). Electroabsorption effects at the band edges of silicon and germanium, *Phys. Rev.*, **145**, 575–583.

6. M. Schmid, M. Kaschel, M. Gollhofer, M. Oehme, J. Werner, E. Kasper and J. Schulze (2012). Franz-Keldysh effect of germanium-on-silicon p-i-n diodes within a wide temperature range, *Thin Solid Films*, **525**, 110–114.

7. M. Oehme, K. Kostecki, M. Schmid, M. Kaschel, M. Gollhofer, K. Ye, D. Widmann, R. Koerner, S. Bechler, E. Kasper and J. Schulze (2014). Franz-Keldysh effect in GeSn pin photodetectors, *Appl. Phys. Lett.*, **104**, 161115.

8. R. S. Jacobsen, K. N. Andersen, P. I. Borel, J. Fage-Pedersen, L. H. Frandsen, O. Hansen, M. Kristensen, A. V. Lavrinenko, G. Moulin, H. Ou, C. Peucheret, B. Zsigri and A. Bjarklev (2006). Strained silicon as a new electro-optic material, *Nature*, **441**, 199–202.

9. Y. Ishikawa, K. Wada, D. D. Cannon, J. Liu, H.-C. Luan and L. C. Kimerling (2003). Strain-induced band gap shrinkage in Ge grown on Si substrate, *Appl. Phys. Lett.*, **82**, 13.

10. D. A. B. Miller, D. S. Chemla, T. C. Damen, A. C. Gossard, W. Wiegmann, T. H. Wood and C. A. Burrus (1984). Quantum-confined Stark effect, *Phys. Rev. Lett.*, **53**, 2173.

11. J. Stark (1913). Splitting of spectral lines of hydrogen in an electric field, *Scientific talks, Nobel Prize in Physics 1919*.

12. Y.-H. Kuo, Y. K. Lee, Y. Ge, S. Ren, J. E. Roth, T. I. Kamins, D. A. B. Miller and J. S. Harris (2005). Strong quantum-confined Stark effect in germanium quantum-well structures on silicon, *Nature*, **437**, 1334–1337.

13. J F. Liu (2016). *Photonics and Electronics with Germanium* (Willey-VCH Verlag).

14. B. Schmidt, Q. Xu, J. Shakya, S. Manipatruni and M. Lipson (2007). Compact electro-optic modulator on silicon-on-insulator substrates using cavities with ultra-small modal volumes, *Opt. Express*, **15**, 3140–3148.

15. X. Tu (2008). Study on the silicon-on-insulator optical waveguide electro-optical modulator, PhD thesis, Institute of Semiconductors, Chinese Academy of Sciences.

16. A. R. M. Zain, N. P. Johnson, M. Sorel and R. M. De La Rue (2008). Ultra high quality factor one dimensional photonic crystal/photonic wire micro-cavities in silicon-on-insulator (SOI), *Opt. Express*, **16**, 12084–12089.

17. Q. Xu, B. Schmidt, S. Pradhan and M. Lipson (2005). Micrometre-scale silicon electro-optic modulator, *Nature*, **435**, 325–328.

18. L. Friedman, R. A. Soref and J. P. Lorenzo (1988). Silicon double-injection electrooptic modulator with junction gate control, *J. Appl. Phys.*, **63**, 1831–1839.

19. A. Cutolo, M. Iodice, P. Spirito and L. Zeni (1997). Silicon electro-optic modulator based on a three terminal device integrated on a low-loss single-mode SOI waveguide, *J. Lightwave Technol.*, **15**, 505–518.

20. A. Liu, R. Jones, L. Liao, D. Samara-Rubio, D. Rubin, O. Cohen, R. Nicolaescu and M. Paniccia (2004). High-speed silicon optical modulator based on a metaloxide-semiconductor capacitor, *Nature*, **427**, 615–618.

21. Q. F. Xu, S. Manipatruni, B. Schmidt, J. Shakya and M. Lipson (2007). 12.5 Gbit/s carrier-injection-based silicon micro-ring modulators, *Opt. Express*, **15**, 430–436.

22. L. Liao, D. Samara-Rubio, M. Morse, A. Liu, D. Hodge, D. Rubin, U. D. Keil and T. Franck (2005). High speed silicon Mach-Zehnder modulator, *Opt. Express*, **13**, 3129–3135.

23. F. Y. Gardes, G. T. Reed, N. G. Emerson and C. E. Png (2005). A submicron depletion-type photonic modulator in silicon on insulator, *Opt. Express*, **13**, 8845–8854.

24. A. Liu, L. Liao, D. Rubin, H. Nguyen, B. Ciftcioglu, Y. Chetrit, N. Izhaky and M. Paniccia (2007). High-speed optical modulation based on carrier depletion in a silicon waveguide, *Opt. Express*, **15**, 660–668.

25. W. M. J. Green, M. J. Rooks, L. Sekaric and Y. A. Vlasov (2007). Ultracompact, low RF power, 10 Gb/s Silicon Mach-Zehnder modulator, *Opt. Express*, **15**, 17106–17113.

26. E. A. J. Marcatili (1969). Dielectric rectangular waveguide and directional coupler for integrated optics, *Bell Syst. Tech. J.*, **48**, 2071–2102.

27. G.-S. Jeong, W. Bae and D.-K. Jeong (2017). Review of CMOS integrated circuit technologies for high-speed photo-detection, *Sensors*, **17**, 1962.

28. P. De Heyn, V. I. Kopp, S. A. Srinivasan, et al. (2017). Ultra-dense 16×56 Gb/s NRZ GeSi EAM-PD arrays coupled to multicore fiber for short-reach 896Gb/s optical links, *Optical Fiber Communication Conference*, Th1B.7.

29. J. Verbist, M. Verplaetse, S. A. Srivinasan, et al. (2017). First real-time 100-Gb/s NRZ-OOK transmission over 2 km with a silicon photonic electro-absorption modulator. *Optical Fiber Communication Conference*, Th5C.4.

30. M. Oehme, E. Kasper and J. Schulze (2013). GeSn heterojunction diode: detector and emitter in one device, *ECS J. Solid State Sci. Technol.*, **2**, 76–78.

Chapter 11

Extension of the Wavelength Regime

The direct bandgap of Ge at 0.8 eV restricts the available wavelength regime in Ge/Si-based photonics from $\lambda = 1.2$ µm (Si absorption) to 1.55 µm (Ge bandgap). Some minor modification allows the Ge direct bandgap to be extended to 1.65 µm, covering both most important telecommunication wavelengths, around 1.3 µm and around 1.55 µm. The minor modification includes thermal mismatch strain (tensile strain of $\varepsilon = 0.2\%$ [1]) and bandgap narrowing from high doping levels [2].

Substantial increase of the upper wavelength limit to beyond 2 µm needs either intraband transition from multiquantum wells (MQWs) [3] or interband transitions from heavily strained Ge [4] or from GeSn alloys [5]. Heavily strained Ge is limited in thickness (at a critical thickness the strain of lattice mismatched layers relaxes [6]). So, GeSn arises as the preferred material to shift the bandgap to lower values than 0.8 eV, which means to higher wavelengths like 1.55 µm.

11.1 Germanium Tin

The research on germanium tin (GeSn) is further stimulated by the predictions of a direct bandgap semiconductor at rather low Sn amounts (around 10%) by prospects for high-speed electronics

---

Silicon-Based Photonics
Erich Kasper and Jinzhong Yu
Copyright © 2020 Jenny Stanford Publishing Pte. Ltd.
ISBN 978-981-4303-24-8 (Hardcover), 978-981-4303-25-5 (eBook)
www.jennystanford.com

(high carrier mobility) and by challenges from material science (nonequilibrium growth for GeSn alloys with Sn above 1%).

11.1.1 Bandgap of GeSn: Indirect to Direct Transition

A rough estimate of the indirect bandgap E_{gL} (L minimum along 111 is the lowest indirect transition from Ge) and the energy difference $\Delta E_g(L/\Gamma)$ between the L minimum and the direct transition (Γ point) may be obtained by a graph showing these variables as functions of $Z^{1/3}$ (Z is the atomic number).

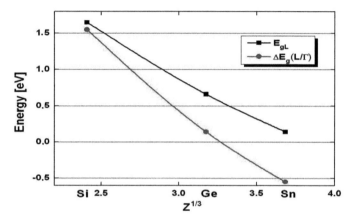

Figure 11.1 Energy values of E_{gL} and $\Delta E_g(L/\Gamma)$ as functions of $Z^{1/3}$.

Both energies show a monotonic behavior with only slight bowing. The $\Delta E_g(L/\tau)$ curve crosses (Fig. 11.1) zero energy (which means direct and indirect transitions have the same energy) at GeSn with rather low Sn amount (about 10%). At higher Sn amounts the semiconductor GeSn is the only bulk group IV semiconductor with a direct bandgap. More sophisticated bandgap theories and optical measurements confirm this predicted transition [7].

The bandgap (Fig. 11.2) as a function of the Sn amount x shows rather strong bowing, with bowing parameters b = 1.94 eV for E_{gdir} and b = 1.23 eV for E_{gL}.

$$E_g(x) = (1-x)E_g(Ge) + x \cdot E_g(Sn) - bx(1-x) \tag{11.1}$$

The indirect to direct crossover is predicted at x = 0.11 and E_g = 0.477 eV by this theory. Later Alberi et al. [8] improved the calculation and she could explain the strong bowing behavior.

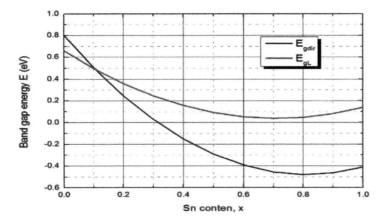

Figure 11.2 Direct E_{gdir} and indirect E_{gL} bandgap of GeSn as functions of Sn amount using data from Ref. [7].

The composition dependence of an alloy's bandgap can be estimated to the first degree by the linear interpolation between the values of the endpoint materials. This linear interpolation is known as generalized Vegard's law (Vegard [9] used it for lattice constant considerations) or, specifically for band properties, as virtual crystal approximation. Band bowing is typically described in a second approximation by a parabolic relationship. Band bowing is primarily caused by constituent mismatch and disorder-related potential fluctuations. While the b values of the bowing parameters in most alloys are smaller than their bandgaps, such as in the case of SiGe ($b > 0.14$), the bowing parameter for GeSn is around twice the gap energy. The electronic structures of highly mismatched alloys have been understood in terms of a band anticrossing model [8]. From this model also assessments of the bowing of band offsets and of the spin-orbit splitting (Δ_0) in the valence band are made.

Type I band alignment between Ge and α-Sn is assumed with $\Delta E_C(\text{Ge/Sn}) = 1$ eV and $\Delta E_V(\text{Ge/Sn}) = 0.2$ eV. (Remark: We count positive energy values for type I alignment). The direct conduction band offset $\Delta E_C(\text{Ge/GeSn})$ of the alloy/Ge interface is then described by

$$\Delta E_C(x) = 1.0 \text{ eV} \cdot x - b_{CB}x(1-x) \tag{11.2}$$

and

$$\Delta E_V(x) = 0.2 \text{ eV} \cdot x - b_{VB}x(1-x). \tag{11.3}$$

Similarly, the valence band offset $\Delta E_V(x)$ is described by Eq. 11.3.

Alberi et al. [8] calculated b_{CB} = 0.7 eV and b_{VB} = 1.24 eV, which exhibit a larger valence band bowing than earlier assessments [10] of b_{CB} = 1.62 eV and b_{VB} = 0.32 eV. The spin-orbit split-off Δ_0 varies rather linearly between the values of Ge (0.3 eV) and α-Sn (0.8 eV).

$$\Delta_0(x) = 0.3 \text{ eV}(1-x) + 0.8 \text{ eV} \cdot x \qquad (11.4)$$

11.1.2 Epitaxial Layer Structures on Silicon

On silicon substrates (bulk Si or silicon on insulator [SOI]) we mainly see the four typical epitaxial configurations as shown in Fig. 11.3.

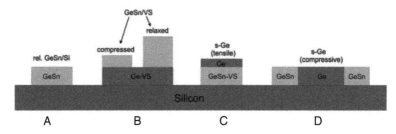

Figure 11.3 Typical GeSn growth structures.

Configuration A shows a straightforward growth of the GeSn layer directly on the Si substrate. The lattice mismatch f is high (linear approximation = 4.3% + 14.7% · x), which results in a strain-relaxed layer already at thicknesses of a few nanometers. In configuration B the GeSn layer is grown on a virtual substrate (Ge VS) composed of a relaxed Ge buffer on Si. The lattice mismatch of low-Sn-content layers relative to the underlying Ge VS is low (f = 14.7% · x), which is comparable with SiGe on Si layers of around 4 × Ge content (e.g., a 5% Sn alloy on a Ge layer is comparable to a 20% Ge alloy on a Si layer). Thin layers are compressively strained, whereas thick layers tend to be strain relaxed. One should consider that the metastable growth regime is large because of the generally low growth temperatures of GeSn alloy layers. For optical properties (direct bandgap) and for electronic properties (high mobility) a tensile strained layer is preferred (configuration C). Here on top of configuration A (relaxed GeSn) a Ge layer or a GeSn layer with lower Sn content is grown (thin enough to be strained). The resulting strain status is a tensile strain. Configuration D is similar for Ge channel transistors to the now used selective source/drain (S/D) regions in p-type Si channel field-effect transistors. The S/D region is grown by selective epitaxy, and it compresses the Ge channel region if GeSn is used for S/D.

The heterostructure has to be combined with a dopant structure for a device application. Let us discuss this with the example of a light-emitting diode (LED) structure as proposed by Sun et al. [11]. Figure 11.4 gives a schematic scheme of the structure, starting from the bottom with a p-type substrate (SOI) and a p-type relaxed buffer from GeSn.

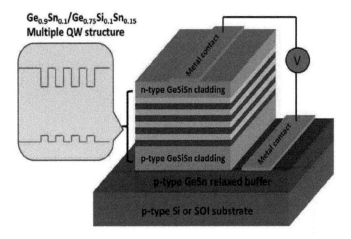

Figure 11.4 Proposal of an LED/laser device structure with active quantum wells from GeSn alloys. Reprinted with permission from Ref. [11] © The Optical Society.

This corresponds to the growth structure A in Fig. 11.3, and it may be described as a virtual GeSn substrate (GeSn VS) as the real substrate is Si but the surface lattice constant is that of GeSn. On this GeSn VS the p-i-n LED follows, which contains a p-type GeSiSn cladding (bottom contact), an intrinsic layer composed of multiple quantum wells (MQWs) from GeSn wells and GeSiSn barriers, and finally an n-type GeSiSn cladding (top contact). The material choice for the claddings and the MQW was such that the strain in the LED was small enough to avoid dislocation generation inside the LED (remember, the GeSn VS contains a misfit dislocation network to allow for lattice accommodation). An equal lattice constant within the MQW was obtained by a combination of binary GeSn and ternary GeSiSn alloy.

11.1.3 Optical and Electrical Results

The basic doping structure is a p-i-n or p/n junction with a voltage-dependent depletion layer at the junction. In the case of metal-semiconductor-metal device structures the highly doped contact layers are replaced by metal layers, which creates a depletion layer (Schottky contact) in the semiconductor.

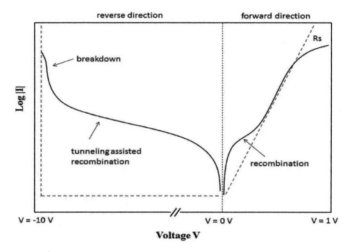

Figure 11.5 Principal scheme of the current characteristics of a junction device in a semilog presentation (log|I|vs. V). Full line: real device; broken line: ideal device with a low recombination center density.

The electrical characterization is important from a practical viewpoint (choice of operating points, noise, and power consumption) and from a characterization viewpoint (rough estimate of the device material quality). Figure 11.5 shows the principal electrical characteristics of a junction device. Ideally (very high quality, very low recombination center density), the forward characterization follows an exponential increase caused by diffusion currents.

$$\ln I = A + B(qV/kT) \qquad (11.5)$$

$$B = 0.4343 = \log(e),$$

which means a decade increase of current I at a 60 mV voltage V increase (T = 300 K). The absolute value of the ideal current is proportional $\exp(-E_g/kT)$. The ideal reverse current is low and constant up to the avalanche breakdown voltage. The real forward characteristics are influenced by the series resistance (damping of the exponential increase at high currents), by the recombination of carriers in the depletion layer (increase of current at a low forward voltage), and by tunneling-assisted recombination (reverse voltage).

In devices on a VS usually recombination currents dominate because the dislocations (threading dislocations cross the depletion layer) are very active recombination centers. Carrier recombination from point defects add to the current because the low growth temperature of GeSn alloys (see next section) creates point defect supersaturating. An example [12] is shown in Fig. 11.6, where the reverse current densities of GeSn p-i-n diodes with different Sn contents are compared with a Ge reference diode.

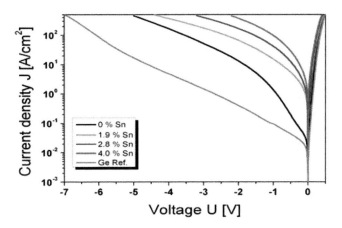

Figure 11.6 Dark current versus voltage for GeSn and Ge photodiodes. Reproduced from Ref. [12], with the permission of AIP Publishing.

The Ge reference diode was grown at 330°C; the tunneling-assisted recombination is mainly caused by the threading dislocations from the Ge VS. The GeSn exhibited higher reverse currents because of the additional point defects generated at the GeSn growth temperature of 100°C. This is easily demonstrated with a Ge p-i-n diode (Sn = 0%) grown at the lower temperature (100°C). The same structure and dislocation density as the Ge reference but a higher reverse current prove a recombination from dislocation and point defects. Material science research is strongly directed to reduce both contributions.

The lower bandgap of GeSn shifts the absorption edge more into the infrared region [13–15]. In a p-i-n junction, that absorption edge can easily be investigated by the photocurrent [12]. The photocurrent near the absorption edge is linearly dependent on the absorption coefficient α and the thickness t, respectively.

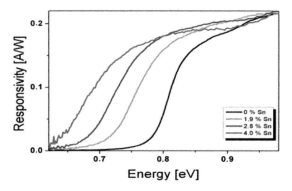

Figure 11.7 Photoresponsivity [12] of GeSn photodiodes with different Sn amounts.

Figure 11.7 demonstrates the infrared shift (lower energy) of the photoresponsivity of GeSn alloys with Sn amounts of a few percent.

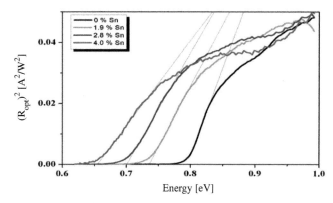

Figure 11.8 Square of the optical responsivity versus energy. A linear increase above the bandgap is typical for direct bandgap transitions. Calculated with data from Fig. 11.7.

The absorption edge is mainly caused by direct transition. Figure 11.8 shows the square of the responsivity, which linearly increases with $(hv - E_g)$ as requested for direct transitions.

The dominant photoluminescence (PL) and electroluminescence (EL) lines originated also from direct transitions [16–18]. Figure 11.9 shows a sequence of PL spectra from GeSn alloys with low Sn content (up to 3%). Already small amounts of Sn cause an observable infrared shift of the PL spectra.

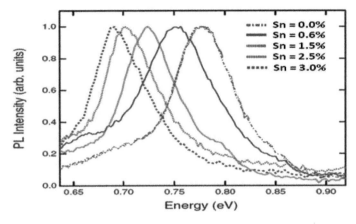

Figure 11.9 Photoluminescence from GeSn on Si. Reproduced from Ref. [17], with the permission of AIP Publishing.

Light emission from GeSn LEDs confirm the redshift of device luminescence [18] with Sn amount. The next figure (Fig. 11.10) compares the emission from a relaxed Ge with that of a compressively strained GeSn (1.3% Sn) LED, both grown on a Ge VS. The GeSn LED shows a direct transition–dominated emission with a clear redshift to a lower emission wavelength. The direct transition dominates in these indirect semiconductors because the direct transition probability is much larger than that of the more populated indirect transition.

A compilation of different methods and authors reveals a rather clear picture for GeSn with low amounts of Sn (up to 7%). The dominant optical absorption and emission process relates to the direct bandgap transition although the semiconductor GeSn is indirect in this low-Sn-content regime. The energy gap decreases (redshift) much more rapidly with Sn content than expected from a linear interpolation, which indicates a strong bowing of the bandgap. The experimental data from the photocurrent [19], PL [17], EL [20], and photoreflection [21] are compared with the theory [8, 22]. They show acceptable agreement within the uncertainties given by the strain in the devices. The dominance of the direct transition made it difficult to measure the indirect-direct band crossover because the weak indirect transition is hard to measure in the thin epitaxial layers. The bowing parameter b describes the parabolic deviation from a linear interpolation. The values for the bowing parameter for the direct transition in GeSn are around 2.5 eV (2.44 eV in Ref. [21], 2.92 eV in Ref. [23], and 2.88 eV in Ref. [24]).

A careful interpretation of experimental observations [22] delivers 7.1% Sn content in strain-relaxed GeSn as a crossover point between direct and indirect bands.

Recent investigations of GeSn alloys with higher Sn concentrations (15%–30% Sn content) demonstrated a clear shift [24] of the absorption edge into the mid-infrared (8 μm wavelength). There are also hints that the parabolic interpolation of the bandgap energy has to be replaced in this higher Sn content regime by a cubic-order equation.

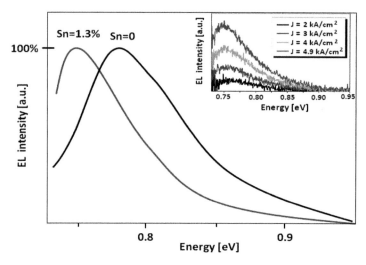

Figure 11.10 Electroluminescence from Ge and GeSn p-i-n junctions on Ge VS (J = 4.9 kA/cm^2). Inset shows current density dependence.

11.1.4 Material Science Challenges

GeSn has been identified as a material of interest for decades [5] for mainly three reasons: (i) the smaller bandgap than Ge extends the infrared range, (ii) a zero bandgap semiconductor may be created with around 40% Sn, and (iii) a direct bandgap transition (the only unstrained group IV direct semiconductor) may be obtained with a rather low Sn amount (8% Sn: expected crossover).

Progress was slow in the past because of serious material science challenges. Only recently, advanced epitaxy methods have pushed the growth of material good enough for device fabrication. The main material challenges stem from:

- Coexistence of a metallic phase and a semiconducting phase (α-Sn) around room temperature. Equilibrium phase transition of the semiconductor/metal takes place at 13°C.
- Large lattice mismatch between α-Sn and Ge (14.7%).
- Large GeSn alloy mixing gap with two-phase equilibrium already there when the Sn amount is more than 1%.
- Low growth temperatures as Sn melts at above 230°C.

Some research groups [21] have overcome the lattice mismatch problem with ternary group III–V compound substrates (Fig. 11.11). The group III–V compounds like GaInAs and GaInSb span the full range of lattice constants for GeSn.

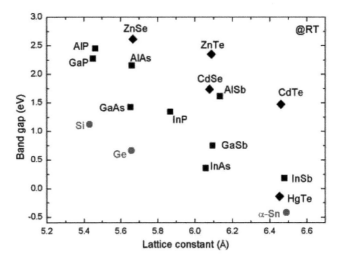

Figure 11.11 Bandgap versus lattice constant of group IV elements and group III–V and group II–VI compounds.

For Si photonics, the low-Sn-content GeSn alloy is addressed with acceptable lattice mismatch values if a VS is used. Even at low Sn amounts, equilibrium growth is not possible, as the phase diagram (Fig. 11.12) shows [5].

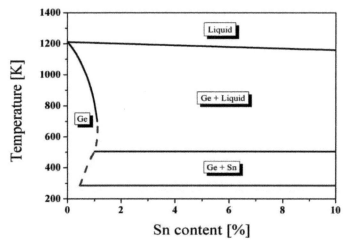

Figure 11.12 Phase diagram of the GeSn material system.

Look, for instance, at 5% Sn content. Then up to 500 K (around 230°C), the GeSn separates into a solid state mixture of Ge and Sn (Ge and Sn are doped to less than 1% by their counterparts); from 230°C to around 900°C, the mixture is solid Ge and liquid Sn droplets; and above 900°C both melt into a GeSn liquid. Modern epitaxy techniques like molecular beam epitaxy and chemical vapor deposition with specific precursors have allowed the epitaxial growth of single-phase GeSn on Si and Ge VS substrates at temperatures low enough (100°C–350°C) to overcome the equilibrium splitting into two phases. Generally, a nonequilibrium supersaturation of point defects (vacancies, interstitials) may occur at these low growth temperatures. Often, annealing steps are introduced during growth to reduce point defect levels and to improve dislocation structure.

11.1.5 Ternary Silicon Germanium Tin Alloys

A wider space in the lattice constant and band offsets may be obtained with ternary alloys. The ternary alloy SiGeSn may gain importance for lattice-matched quantum well structures and mid-infrared (0.2–0.6 eV) devices. An overview about calculation of the bandgaps and their photonic applications is given in Ref. [3]. The lattice constant a of the ternary alloy $Si_x Ge_{1-x-y} Sn_y$ is given by

$$a = a_{Si} \cdot x + a_{Ge}(1 - x - y) + a_{Sn} \cdot y$$

$$+ b'_{SiGe} x(1-x) b'_{SnGe} y(1-y), \qquad (11.6)$$

The lattice constants of Si, Ge, and Sn were taken as 0.54307 nm, 0.56575 nm, and 0.64912 nm, respectively. The lattice bowing parameters b' were chosen as $b'_{SiGe} = -0.026$ and $b'_{GeSn} = -0.166$, while the bowing of SiSn was taken as zero.

The most interesting applications of ternary alloys concern the possibility of creating lattice-matched quantum wells with practically usable band offsets. In Fig. 11.13 the direct conduction band energy (Γ) and the valence band edge E_v are shown as functions of the Sn content for a lattice-matched GeSiSn/GeSn heterojunction [25].

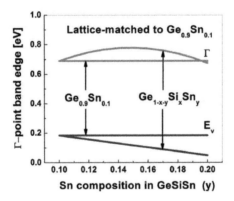

Figure 11.13 Direct conduction band edge and valence band edge of a lattice-matched Ge SiSn/Ge$_{0.9}$Sn$_{0.1}$ heterojunction [25].

As seen in Fig. 11.13 the ternary alloy may be used as a barrier material with around 100 meV energy barrier to the GeSn well.

11.2 Tensile-Strained Germanium

Generally, strain shifts the energies of the valence edges and band and the conduction band may split into sub-bands from otherwise degenerate states. Such degenerate states are light and heavy holes (lh and hh) or the fourfold degenerate indirect L valleys in the conduction band. We consider now the biaxial stress of a GeSn layer on a (001) substrate. Biaxial stress leads to a tetragonal distortion of the strained lattice cell with equal strain ε in the in-plane directions (x and y) and an opposite perpendicular strain $\varepsilon_{\text{perp}}$ (z direction).

$$\varepsilon_{\text{perp}} = -\varepsilon \frac{2\upsilon}{1-\upsilon} \qquad (11.7)$$

Here υ is the Poisson ratio, which is assumed to be equal to $\upsilon = 0.27$.

Tetragonal distortion of the lattice cell may be considered as a sum of volume change ($2\varepsilon + \varepsilon_{\text{perp}}$) by all-around compression and a uniaxial extension in the perpendicular direction ($\varepsilon_{\text{perp}} - \varepsilon$). This two-term representation of biaxial stress is performed because all fundamental measurements of strained materials are done by volume change in a high-pressure cell and by uniaxial dilatation of a material rod. The volume change of a biaxial stressed epitaxial layer is given by (the numbers are calculated with $\upsilon = 0.27$)

$$2\varepsilon + \varepsilon_{\text{perp}} = 2\left(\frac{1-2\upsilon}{1-\upsilon}\right)\varepsilon = 1.26\varepsilon \qquad (11.8)$$

and the uniaxial strain along the z direction is given by

$$\varepsilon_{\text{perp}} - \varepsilon = -\frac{1-\upsilon}{1+\upsilon}\varepsilon = -1.74\varepsilon. \qquad (11.9)$$

The hydrostatic strain component (Eq. 11.8) causes an energy shift of the valence and conduction bands; the difference between both is measured as a bandgap change. For an overview on general strain effects, see Ref. [21]. The change in the bandgap energy is proportional to the hydrostatic deformation potential parameter a_{g}. The direct transition deformation potential values for the coefficient $a_{\text{g}}(\Gamma)$ vary around -10 eV (+/- 20%), as discussed in more detail in Ref. [26]. The coefficients $a_{\text{g}}(L)$ for the indirect transition are also negative but much smaller (about -2.5 eV), with a relative uncertainty of more than 20%.

The large variations in deformation potential values result in uncertainties regarding bandgap extraction from strained materials. In the future biaxial stressed epitaxy layers may be used as a test material for the refinement of deformation potential measurements, but it is crucial to use well-defined structure parameters.

The uniaxial component (Eq. 11.9) causes a splitting of the valence band edge for the hh and lh bands. There is no splitting in the conduction band because the L valleys in the <111> direction are symmetric to the substrate orientation. This is simpler than in the SiGe/Si case, where the dominating indirect X valleys are split in the in-plane and perpendicular conduction sub-bands. The uniaxial deformation potential value b_g for the (100) substrate orientation amounts to about –2.5 eV. The deformation potential value b_g is the same for direct and indirect bands in the given orientation because both transitions address the same valence sub-bands, lh or hh.

The relation between strain and the lh bandgap is slightly nonlinear. The reason is given by the interaction with the third valence sub-band (spin-orbit split-off band), which is not considered here because its energy levels are lower by the split orbit energy difference $\Delta_0 = 0.287$ eV. Combining Eqs. 11.8 and 11.9, we obtain for biaxially stressed Ge the linear relations for the direct transitions to the lh and hh valence sub-bands. These linear relations may be used up to 1% strain; for higher values the nonlinear interaction with the split-off valence band has to be taken into account [26].

$$E_g(\text{lh}) = E_g(0) + \varepsilon(1.26a + 1.74b) \tag{11.10}$$

$$E_g(\text{hh}) = E_g(0) + \varepsilon(1.26a - 1.74b) \tag{11.11}$$

For compressive strain (negative value of ε), the hh bandgap is the lower one (Eq. 11.11) and this will dominate the bandgap emissions. The direct transition (hh) is clearly above the unstrained energy value $E_g(0)$ because the large deformation potential parameter a dominates. For indirect transition (hh) the energy is roughly strain independent because the smaller a coefficient contribution is partly cancelled by the b term. The compressive biaxial stress enhances the indirect character of Ge; it is only interesting for adjusting the modulation wavelengths of electroabsorption modulators.

The most important stress situation in Si-based photonics is given by tensile biaxial stress (Eq. 11.10). The energy of the direct transition decreases faster than that of the indirect transition, yielding a crossover between indirect/direct conduction bands at about 2% in-plane strain. The band energy decreases to about 0.5 eV at 2% in-plane strain, resulting in a 2.5 μm emission wavelength for the crossover situation. The crossover with uniaxial stress needs much higher strain levels (about 5%).

11.3   Conclusions

Application of strain and alloying of Ge with Sn are the prime candidates for the extension of the wavelengths in Si-based photonics beyond 1.6 μm toward the midinfrared regime. Biaxial in-plane stress leads to a tetragonal distortion of the lattice cell. Tensile strain favors reduction of the energy separation between direct band and indirect band (136 meV in unstrained Ge), and it reduces the bandgap to 0.5 eV for 2% biaxial tensile strain. For this biaxial strain value, an indirect-direct crossover takes place in the semiconductor. Uniaxial strain is less effective, requiring 5% for the conduction band crossover.

GeSn is a promising small-bandgap group IV semiconductor. Infrared shift of absorption, direct luminescence, and reflection are proven experimentally (extended infrared window for silicon-based photonics). First fundamental devices (photodetector, LED, optically stimulated lasers) are demonstrated with infrared light emission and detection up to 3 μm wavelength. Direct-indirect crossover takes place for strain-relaxed GeSn with about 7%–9% Sn amount. The crossover GeSn shows a similar bandgap reduction (to 0.5 eV) as the strained Ge.

Advanced metastable material preparation paved ways to direct group IV semiconductors for midinfrared (>2.5 μm) silicon-based photonics. The groups of John Kouvetakis and Jose Menendez [24] demonstrated the growth of GeSn layers with high Sn contents (13%–27%) directly on Si. The high lattice mismatch of GeSn on Si forced a rapid mismatch accommodation at the interface, resulting in layers with a rather small residual strain (between –0.5% and +0.2%). The cut-off wavelengths of these layers shifted from 3 μm to 8 μm with increasing Sn content.

Material science and fabrication fights with nonequilibrium alloy formation, lattice mismatch accommodation, strain management, defect creation from dislocation networks, and ultralow temperature processing, but steep progress is expected.

References

1. X. Sun, J. Liu, L. C. Kimerling and J. Michel (2009). Room-temperature direct bandgap electroluminesence from Ge-on-Si light-emitting diodes, *Opt. Lett.*, **34**, 1198.

2. E. Kasper, M. Oehme, J. Werner, T. Aguirov and M. Kittler (2012). Direct band gap luminescence from Ge on Si pin diodes, *Front. Optoelectron.*, **5**, 256.

3. P. Moontragoon, R. A. Soref and Z. Ikonic (2012). The direct and indirect bandgaps of unstrained $Si_xGe_{1-x-y}Sn_y$ and their photonic device applications, *J. Appl. Phys.*, **112**, 073106.

4. P. Boucaud, M. El Kurdi and J. M. Hartmann (2004). Photoluminescence of a tensilely strained silicon quantum well on a relaxed SiGe buffer layer, *Appl. Phys. Lett.*, **85**, 46.

5. E. Kasper, J. Werner, M. Oehme, S. Escoubas, N. Burle and J. Schulze (2012). Growth of silicon based germanium tin alloys, *Thin Solid Films*, **520**, 3195.

6. E. Kasper, N. Burle, S. Escoubas, J. Werner, M. Oehme and K. Lyutovic (2012). Strain relaxation of metastable SiGe/Si: investigation with two complementary X-ray techniques, *J. Appl. Phys.*, **111**, 063507.

7. V. R. D'Costa, C. S. Cook, A. G. Birdwell, C. L. Littler, M. Canonico, S. Zollner, J. Kouvetakis and J. Menendez (2006). Optical critical points of thin-film $Ge_{1-y}Sn_y$ alloys: a comparative $Ge_{1-y}Sn_y/Ge_{1-x}Si_x$ study, *Phys. Rev. B*, **73**, 125207.

8. S. Alberi, J. Blacksberg, L. D. Bell, S. Nikzad, K. M. Yu, O. D. Dubon and W. Walukiewicz (2008). Band anticrossing in highly mismatched Sn_xGe_{1-x} semiconducting alloys, *Phys. Rev. B*, **77**, 073202.

9. L. Vegard (1921). Die Konstitution der Mischkristalle und die Raumfüllung der Atome, *Z. Phys.* (in German) **5**, 17–26.

10. R. S. Bauer and G. Margaritondo (1987). Probing semiconductor/ semiconductor interfaces, *Phys. Today*, **40**, 27.

11. G. Sun, R. A. Soref and H. H. Cheng (2010). Design of a Si-based lattice-matched room-temperature GeSn/GeSiSn multi-quantum-well mid-infrared laser diode, *Opt. Express*, **18**, 19957.

12. M. Oehme, M. Schmid, M. Kaschel, M. Gollhofer, D. Widmann, E. Kasper and J. Schulze (2012). GeSn p-i-n detectors integrated on Si with up to 4% Sn, *Appl. Phys. Lett.*, **101**, 141110.

13. J. Mathews, R. Roucka, J. Xie, S.-Q. Yu, J. Menendez and J. Kouvetakis (2009). Extended performance GeSn/Si(100) p-i-n photodetectors for

full spectral range telecommunication applications, *Appl. Phys. Lett.*, **95**, 133506.

14. S. Su, B. Cheng, C. Xue, W. Wang, Q. Cao, H. Xue W. Hu, G. Zhang, Y. Zuo and Q. Wang (2011). GeSn p-i-n photodetector for all telecommunication bands detection, *Opt. Express*, **19**, 6400.

15. J. Werner, M. Oehme, M. Schmid, M. Kaschel, A. Schirmer, E. Kasper and J. Schulze (2011). Germanium-tin p-i-n photodetectors integrated on silicon grown by molecular beam epitaxy, *Appl. Phys. Lett.*, **98**, 061108.

16. R. Roucka, J. Mathews, R. Beeler, J. Tolle, J. Kouvetakis and J. Menendez (2011). Direct gap electroluminescence from Si/Ge1−ySny p-i-n heterostructure diodes, *Appl. Phys. Lett.*, **98**, 061109.

17. J. Matthews, R. Roucka, J. Xie, S.-Q. Yu, J. Menendez and J. Kouvetakis (2009). Extended performance GeSn/Si(100) p-i-n photodetectors for full spectral range telecommunication applications, *Appl. Phys. Lett.*, **95**, 133506.

18. M. Oehme, J. Werner, M. Gollhofer, M. Schmid, M. Kaschel, E. Kasper and J. Schulze (2011). Room-temperature electroluminescence from GeSn light-emitting pin diodes on Si, *IEEE Photonics Technol. Lett.*, **23**, 1751–1753.

19. M. Oehme, E. Kasper and J. Schulze (2012). GeSn photodetection and electroluminescence devices on Si, *ECS Trans.*, **50**(9), 583–590.

20. T. Arguirov (2014). Pers. Communication.

21. H. Lin, R. Chen, W. Lu, Y. Huo, T. I. Kamins and J. S. Harris (2012). Investigation of the direct band gaps in $Ge_{1-x}Sn_x$ alloys with strain control by photoreflectance spectroscopy, *Appl. Phys. Lett.*, **100**, 102109.

22. S.-Q. Liu and S.-T. Yen (2019). Extraction of eight-band k · p parameters from empirical pseudopotentials for GeSn, *J. Appl. Phys.*, **125**, 245701.

23. H. Tran, W. Du, S. A. Ghetmiri, A. Mosleh, G. Sun, R. A. Soref, J. Margetis, J. Tolle, B. Li, H. A. Naseem and S.-Q. Yu (2016). Systematic study of $Ge_{1-x}Sn_x$ absorption coefficient and refractive index for the device applications of Si-based optoelectronics, *J. Appl. Phys.*, **119**, 103106.

24. C. Xu, P. M. Wallace, D. A. Ringwala, S. L. Y. Chang, C. D. Poweleit, J. Kouvetakis and J. Menéndez (2019). Mid-infrared (3-8 μm) $Ge_{1-y}Sn_y$ alloys (0.15 < y < 0.30): synthesis, structural, and optical properties, *Appl. Phys. Lett.*, **114**, 212104.

25. G. Sun (2011). ICSI-7, Invited Talk.

26. E. Kasper and M. Oehme (2015). Germanium tin light emitters on silicon, *Jpn. J. Appl. Phys.*, **54**, 04DG11.

Chapter 12

Laser

Coherent light emission from lasers is considered as especially valuable because of phase information, good collimation, and high intensity.

Lasers are based on self-oscillations of light emitting systems through stimulated emission. To gain enough to overcome absorption, inversion of the quantum states is needed between which the laser transition occurs. Concentration of the emitted photons in only a few modes is obtained through forward and back reflections on mirrors. Photon emission is increased as the stimulated emission probability is proportional to the number of photons per mode.

For silicon-based photonic system different geometrical configurations of lasers are possible. The three most important ones are summarized in Table 12.1.

Often, monolithically integrated lasers on Si are considered essential for Si-based photonic systems. This is only true if a large number of light sources on a chip is required and if the cost advantages of wafer manufacturing may be realized for high-volume market segments. Nevertheless, monolithic integration of lasers on Si is considered as the "holy grail" of Si photonics as this allows the most flexible solution of complex optical sensors and communications systems. The technical goal on a long-term scale

---

Silicon-Based Photonics
Erich Kasper and Jinzhong Yu
Copyright © 2020 Jenny Stanford Publishing Pte. Ltd.
ISBN 978-981-4303-24-8 (Hardcover), 978-981-4303-25-5 (eBook)
www.jennystanford.com

270 | *Laser*

is given by monolithic integration of high-efficiency lasers where strain engineering and Sn alloy engineering of Ge on Si lasers seem to be the most promising solution paths. High efficiency of power conversion is the key for integrated lasers. Otherwise, they are unwanted hot spots on already thermally stressed chips.

Table 12.1 Laser configurations in Si-based photonic systems

Laser	Light coupling from external source	Hybrid laser mounting	Monolithically integrated laser
Advantage	High power, heat outside chip	High efficiency, flexibility of laser positions	Large number of lasers, wafer manufacturing
Drawback	High effort for chip housing	Hot spots on chip	Low efficiency, wavelength limitations
Application potential	Clock distribution on processor chips	Fast processor memory connections	Full system on a chip
Availability	Technical solution available	Research demonstration	First research results

Different configurations of lasers are required for more near-term solutions and for chips with a low number of light sources.

Light coupling from external sources is an already available technical solution, which will continue to be the preferred configuration if the heat from high-power lasers is left outside the densely packed chip and if the laser light may be split and distributed around the chip dimension by a photonic waveguide network. This configuration has high application potential for optical clock distribution on processor chips. Optical clock distribution promises clear advantages beyond 10 GHz clock frequency. The topology of on optical clock is compared to an electrical clock in Table 12.2.

The optical clock needs electro-optical converters, at the source side the modulator and at the drain side the photodiode. The speed of the optical clock is limited by the electro-optical converter devices, which are now around 50 GHz fast but have the potential for more than 100 GHz speed.

Table 12.2 Optical clock compared to an electrical clock distribution on a processor chip

	Electrical clock	**Optical clock**
Clock source	Clock generator	Laser and modulator
Source-processor connection	Electrical interconnect to processor pins	Optical fiber to processor housing
Processor chip input	P-i-n connection to the upper level of chip metallization	Fiber output to tapered grating coupler
Clock signal distribution	Upper level of metallisation; RC time limited	Photonic waveguide lines and splitters
Transistor clock	Electrical connection from the upper level to the lower level of metallization	Electro-optical conversion by photodetectors; short-distance connection by lower-level metal lines

12.1 Silicon-Based Laser Approaches

There are many different laser realizations, but in semiconductors band-to-band transitions are the preferred ones for lasers in the visible or near-infrared regime. Many direct semiconductors have proven to be excellent laser materials in diode lasers, where carrier injection is provided from forward-biased p/n junctions.

Silicon, however, is an indirect semiconductor with the lowest direct transition (more than 2 eV) above the indirect one. Occupation inversion of this high-lying direct band is practically impossible, which cancels bulk silicon from the list of diode laser materials. The optical properties of crystalline bulk silicon were shown to be modified by disorder (amorphous Si [α-Si]), size (nanocrystal Si [nc-Si]) [1], or periodicity modulations (superlattice [SL]) [2]. Nanocrystals are rather easily produced by etching to porous silicon [3] or by annealing of Si-rich oxide [4]. Efficient room temperature visible photoluminescence from porous Si [3] and net optical gain with optical stimulation from Si nanocrystals [4] embedded in SiO_2 prove the successful modification of Si optical properties by size effects. These size effects are a combination of wavefunction localization, energy-level quantization, and partly binding effects on nc-Si surfaces.

In 1974, Gnutzmann and Clausecker [5] made the theoretical prediction that the imposition of a proper new periodicity in an indirect-gap semiconductor will fold the bands into a smaller Brillouin zone (BZ). This could bring the minimum of the conduction band to the Γ point of the BZ, thus producing a quasi-direct-gap material. The band folding concept was convincingly proven with observation of folded phonons in Si/Ge superlattices using Raman spectroscopy [6]. The first experimental evidence of photoluminescence from zone-folded states of superlattices was given by Zachai et al. [7].

Electroluminescence from p-i-n diodes was obtained with the Si/Ge SL in the intrinsic part of the diode. Light emission could be demonstrated both from mesa type and waveguide devices [8].

Despite these and other exciting research results a laser could not be realized from structurally modified silicon, which switched the interest of the device- and circuit-oriented photonics community to a more practical solution path: hybrid integration of silicon photonics with the well-established group III/V laser technology [9]. As suggested in Ref. [8] two hybrid methods are preferred in terms of technology compatibility and scalability. The first one is hybrid laser integration through evanescent coupling with the photonic silicon-on-insulator (SOI) wafer. In this method a shared waveguide is created between the III/V laser and silicon. Evanescent coupling is used to deliver optical energy into the SOI photonic platform. This approach needs close proximity (<100 nm) between the III/V and Si material. This close proximity can be obtained by top-down bonding of the III/V laser on the SOI waveguide structure [10]. The other favorable approach is a special form of epitaxial transfer called transfer printing [8], because arrays of epitaxial layer coupons are collectively transferred to the receiving substrate using a specific stamp.

The success of hybrid laser integration had a large influence on the acceptance of silicon photonics but also inspired new research on monolithic integration approaches. Two monolithic approaches emerged as promising variants. The long-term approach is based on the metastable GeSn alloy that is predicted to be a direct semiconductor containing more than 10% Sn. We have dedicated a separate chapter to the GeSn alloy because of its potential importance not only for Si-based monolithic lasers but also for the shift into the mid-infrared-wavelength regime. A more near-term solution is based on the indirect semiconductor material Ge for which optically stimulated laser [11] and electrically stimulated laser operation at room temperature [12] have already been reported.

In the next section we review our basic knowledge about semiconductor lasers as obtained by more than 30 years of laser improvement in the III/V field [13–17]. The reader familiar with these principles may skip this section. Then we discuss why another indirect semiconductor (Ge) is, in contrast to Si, a reasonable candidate for laser operation and we give the status of experimental verification.

12.2 Basic Laser Physics

There are three main interactions that light can have with semi-conductors. In an earlier chapter the absorption process of a photon has been discussed with regard to photodetectors. Figure 12.1a shows the absorption process in a schematic two-level diagram where a carrier is excited by the photon from a group state E_0 to an excited state E_1 if the photon has a frequency $v = E/h$, with $E = E_1 - E_0$.

12.2.1 Spontaneous Emission in a Two-Level System

For a carrier sitting in an upper energy level (Fig. 12.1b) this state is unstable. At a specific time without any external influences it will make a transition to the ground state, emitting a photon of frequency given by $hv = E_1 - E_0$ This is termed spontaneous emission and is the emission process in light-emitting diodes (LEDs). If a photon of energy hv is incident on the carrier in the upper energy state E_1 then this photon can stimulate the emission of another photon of the same frequency and phase (i.e., it is coherent with the first photon), with the carrier making a transition to the ground state (E_0). This is the stimulated emission process in a laser. The radiation is monochromatic because all the photons have energy hv, and it is coherent because all the photons are in phase.

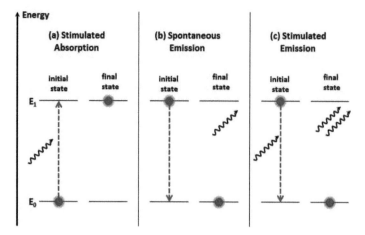

Figure 12.1 A schematic diagram of the photonic transitions between two energy states. (a) When a carrier in the lower level absorbs a photon. (b) If a carrier is in the upper level then it can fall to the lower energy level emitting a photon with frequency $v = E/h$, where E is the energy difference between the two levels. (c) If a carrier is in the upper level it can be stimulated to fall to the lower level by another photon of frequency $v = E/h$, emitting a second photon of the same frequency that is coherent (in phase) with the first photon.

Let us assume that the instantaneous populations of the states E_0 and E_1 are n_0 and n_1 respectively. Under thermal equilibrium and for $(E_1 - E_0) > 3 k_B T$ the population of the states is given by the Boltzmann distribution.

$$\frac{n_1}{n_0} = \exp\left(-\frac{E_1 - E_0}{k_B T}\right) = \exp\left(-\frac{hv}{k_B T}\right). \quad (12.1)$$

As is consistent with a Boltzmann distribution, there are more carriers in the low-energy states. In the steady state, the stimulated emission rate and the spontaneous emission rate must be balanced by the rate of absorption to maintain the populations n_0 and n_1 constant.

The stimulated emission rate is proportional to the photon-field energy density in the system defined as $I(hv)$. The stimulated emission rate can, therefore, be written as $Bn_1 I(hv)$. The spontaneous emission rate from E_0 to E_1 is defined as An_1, with A a constant and independent of the photon density. A and B are named the Einstein coefficients. The absorption rate is proportional to the carrier population in the lower level and the photon-field energy density. Hence it is given by $Bn_0I(hv)$. In the steady state we have stimulated emission rate + spontaneous emission rate = absorption rate, that is

$$Bn_1I(hv) + An_1 = Bn_0I(hv). \tag{12.2}$$

For a laser it is stimulated emission that we are interested in dominating over spontaneous emission. Therefore, rewriting Eq. 12.2 as

$$\frac{\text{Stimulated emission rate}}{\text{Spontaneous emission rate}} = \frac{B}{A}I(hv). \tag{12.3}$$

it is clear that to enhance the stimulated emission, the photon-field density requires to be very large.

Equation 12.2 can also be rearranged to give

$$\frac{\text{Stimulated emission rate}}{\text{Absorption rate}} = \frac{n_1}{n_0}. \tag{12.4}$$

Therefore, for the stimulated emission to dominate over the absorption of photons, the upper energy state, E_1 requires a larger population of states than the lower energy state, E_0 that is, population inversion. Population inversion demands nonequilibrium caused by the injection of carriers.

12.3   Optical Gain in a Semiconductor

Light will be amplified in semiconductors similar to a two-level system if enough electrons are in the conduction band and enough holes in the valence band that stimulated emission overwhelm absorption. Population inversion may be obtained by absorption of a pumping laser or by injection of carriers from a forward-biased diode structure. Electrically stimulated lasers have the higher practical importance in semiconductors.

The easiest way to enhance the photon-field density is to produce a resonant optical cavity. The simplest type of cavity is a Fabry–Perot ridge waveguide (Fig. 12.2), where a ridge is etched out of the active semiconductor material. The refractive index change between the semiconductor and air is enough to reflect around 30% of the radiation at each facet.

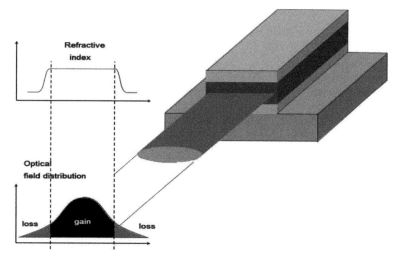

Figure 12.2 Modal confinement into a ridge waveguide using a change in refractive index between the semiconductor and air.

Hence along the length of the cavity, standing wave modes are set up such that the frequency $v = kc/2L$, where L is the length of the cavity, c is the speed of light in the medium, and k is the number describing the order of resonance. Once a cavity has been fabricated and an active semiconductor structure with population inversion has been placed in the cavity, the cavity modes (Fig. 12.3b) will convolve with the broadened electroluminescence spectrum of the transition E_1 to E_0 (Fig. 12.3a), forming the laser spectrum shown in Fig. 12.3c. This is a Fabry–Perot cavity mode spectrum, with several sharp peaks corresponding to the modes along the length of the cavity.

A diode on which a forward bias is applied shows a current that exponentially increases with applied bias voltage. The number of injected carriers increases proportionally to the current if we assume a concentration-independent lifetime of the carriers. Current has to be high enough to generate a positive net gain, which means the semiconductor gain has to overcome the losses by absorption and scattering. The current density at which lasing starts is called threshold current density J_{th} (Fig. 12.4).

Optical Gain in a Semiconductor | 277

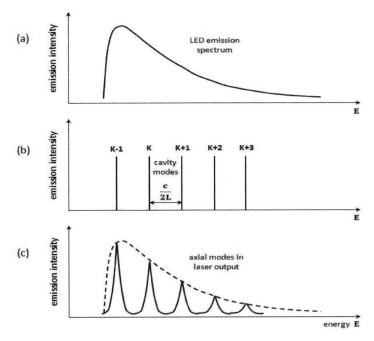

Figure 12.3 Photoemission intensity as a function of energy. In (a) the spontaneous emission (LED) is shown. In (b) the cavity modes of a Fabry–Perot resonator are shown. In (c) the stimulated emission of a laser with axial modes from the resonator is shown.

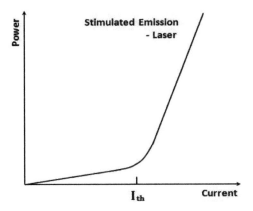

Figure 12.4 Emission power of an electrically stimulated laser as a function of the current. Laser operation sets in from a threshold current I_{th}.

The energy states in semiconductors are grouped in bands. Transitions are allowed for photon energies larger than the bandgap. Semiconductors show a frequency-dependent gain spectrum if population inversion is obtained.

The principle of frequency-dependent gain is explained in the example of a semiconductor with a parabolic energy–momentum relation, which means a direct semiconductor within the effective mass model.

Figure 12.5 shows the energy states and the optical transitions in the active region of a diode laser, where carrier injection from n- and p-side increases carrier levels above equilibrium. In equilibrium the carrier occupation is described by Fermi–Dirac statistics, with a unique Fermi energy level F for electrons and holes. In nonequilibrium, the carrier distribution within the bands is described again by Fermi–Dirac statistics but with separate Fermi energy levels F_C and F_V for the conduction band and the valence band, respectively, if intraband energy relaxation is much faster than band-to-band transitions, a situation we would assume to occur.

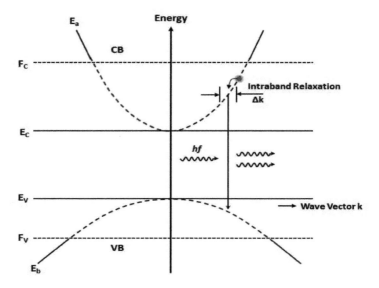

Figure 12.5 Parabolic bandgap relation for energy E versus wave vector amount $\bar{k}$ E_c and E_V band edges of the conduction band and the valence band. Nonequilibrium is described by different quasi-Fermi energy levels F_c and F_v for the both bands, respectively.

Optical transitions in the energy-momentum presentation of Fig. 12.5 are approximately vertical, which means without an essential change in momentum $\bar{k}$, because the momentum $\bar{p}$ of a photon is rather small.

We start with the gain spectrum definition for a two-level system

$$g(v) = \sigma(n_1 - n_0), \tag{12.5}$$

where σ is the cross section of stimulated emission/absorption. It is easier in a semiconductor to start with a momentum-based description and then transfer to a frequency-dependent gain. The density of states (DOS) $Z(k)$ is given in a bulk semiconductor by

$$Z(k) = k^2/\pi^2, \tag{12.6}$$

which gives for $(n_1 - n_0)$ in a nonequilibrium semiconductor

$$n_1 - n_0 = \frac{k^2}{\pi^2} \Delta k (f_C - f_V), \tag{12.7}$$

with f_C and f_V the occupation probabilities for electrons and holes, respectively, from Fermi–Dirac statistics.

Returning to a frequency picture, we have

$$\Delta k = \frac{dk}{dE} \Delta E = \frac{dk}{dv} \Delta v \tag{12.8}$$

and

$$g(v) = \sigma(n_2 - n_1) = \sigma \frac{k^2}{\pi^2} \frac{dk}{dv} \Delta v (f_C - f_V). \tag{12.9}$$

The relation dk/dv for parabolic bands is extracted from

$$E_a - E_C = \frac{\hbar^2 k^2}{2m_C}; \quad E_b - E_V = -\frac{\hbar^2 k^2}{2m_V}, \tag{12.10}$$

where the effective masses of electrons and holes are m_C and m_V, respectively, and E_a and E_b are the energy levels in the conduction band and the valence band, respectively, which correspond to the transition of energy hv:

$$hv = E_g + \frac{\hbar^2 k^2}{2} \left(\frac{1}{m_C} + \frac{1}{m_V} \right) = E_g + \frac{\hbar^2 k^2}{2m_r}, \tag{12.11}$$

where m_r is the reduced effective mass of the electron-hole pair. The optical transition probability between bands of different effective masses is proportional to $m_r^{3/2}$, which is known as the joint-density-of-states model.

$$\frac{1}{m_r} = \frac{1}{m_C} + \frac{1}{m_V} \tag{12.12}$$

Replacing k by v using the effective mass relations gives finally for the gain spectrum $g(v)$

$$g(v) = \frac{\sigma}{\hbar^2 \pi}(2m_r)^{3/2}(h v - E_g)^{1/2}(f_C - f_V)\Delta v. \tag{12.13}$$

The state occupation probabilities f_C and f_V are given as

$$f_C = \frac{1}{\left\{\left[\exp\frac{(E_a - F_C)}{kT}\right] + 1\right\}} \tag{12.14}$$

and

$$f_V = \frac{1}{\left\{\left[\exp\frac{(E_b - F_V)}{kT}\right] + 1\right\}}, \tag{12.15}$$

with F_C and F_V the quasi-Fermi energy levels of a nonequilibrium semiconductor. The corresponding energy levels E_a and E_b in conduction and valence bands may be expressed as a function of the photon energy hv:

$$E_a = (h v - E_g)\frac{m_r}{m_C} + E_C \tag{12.16}$$

$$E_b = (h v - E_g)\frac{m_r}{m_V} + E_V \tag{12.17}$$

12.4 Semiconductor Heterostructure Laser

In our basic consideration of the laser operation a p-i-n diode junction was used for injection of the high current that allows inversion of the quantum states. This means that one has to fill the conduction band states at least a few kT. The development of direct bandgap lasers from III/V material demonstrated that band filling could be effectively obtained by carrier confinement in semiconductor heterostructures [13]. A material sequence ABA, where material B has a lower energy for carriers than material A, is called a double heterostructure (DH), which collects the injected carriers within the material slab B. The DOS $Z(E)$ for these collected carriers is bulk like (3D movement) if the thickness d of the slab B is on the order of the intrinsic region of the doping structure (typically 150 nm). Note that the DOS is usually given as a function of energy E. The momentum-based DOS $Z(k)$ of Eq. 12.6 may be transferred to $Z(E)$ by using Eq. 12.8. Quantum size effects [14] will have a considerable influence on the DOS (Fig. 12.6) if the thickness d shrinks to several tens of nanometers. The structure is called quantum film or quantum well (QW) if the carrier confinement is in one direction (2D movement), and it is called quantum wire or quantum box if confinement is performed in two directions (1D movement) or all three directions (0D movement), respectively. A scheme [15] of the quantum size structures and their DOS $Z(E)$ is given in Fig. 12.6. A DH laser and a QW laser differ only in their thickness d, which changes not only the DOS but also the optical confinement.

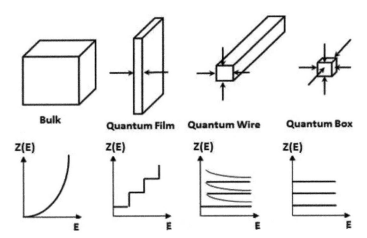

Figure 12.6 Schematics of 3D, 2D, 1D, and 0D quantum systems and corresponding densities of states Z(E). Graph compiled using data from Ref. [15].

The wavelength of the infrared light is only about several hundred nanometers in a semiconductor because of the high refractive index n. The thickness d of a DH laser is comparable with the wavelength of light; in contrast the thickness of QW is much smaller than the wavelength. This is discussed later, under optical confinement factor.

A main parameter is the optical confinement factor Γ, which represents the fraction of the optical wave in the active layer material, that is, the efficiency of an emitted photon to interact with another electron–hole pair in order to further induce stimulated emission. This confinement factor depends on the active layer thickness and on the difference in the indexes of refraction between the active layer and confining materials.

$$\Gamma = \frac{\int_{-d/2}^{+d/2} |E(z)|^2 \, dz}{\int_{-\infty}^{+\infty} |E(z)|^2 \, dz} \quad (12.18)$$

As indicated in Fig. 12.7, the d^2 variation of Γ for simple heterostructures [13] leads to extremely small values of Γ in QWs, typically $4 \cdot 10^{-3}$ for $d = 10$ nm. Using the separate-confinement heterostructure (SCH) scheme, with a fixed cavity to confine the optical wave separately from the electron wave (QW), one obtains $\Gamma \approx 3 \cdot 10^{-2}$ in a 10 nm GaAs/GaAlAs QW. Note that the width of the single quantum well (SQW) as the confining waveguide is too thin whereas in the SCH case, the optical waveform is independent of the QW thickness as it is confined by the intermediate composition layer.

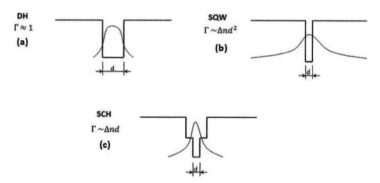

Figure 12.7 Optical confinement factors in double heterostructures (DHs). Schematics of conduction band edge and corresponding optical wave for a DH (a), a single quantum well (SQW) (b), and a separate-confinement heterostructure (SCH) (c).

The gain g_{th} at the threshold is obtained by stating that the optical wave intensity after a roundtrip in the cavity must stay equal under the opposite actions of losses and gain. This is conveniently written as:

$$I_0 R^2 \, e^2 (\Gamma g_{th} - \alpha) L = I_0, \qquad (12.19)$$

where R is the facet reflectivity; L is the laser length; Γg_{th} represents the modal gain per unit length of the optical wave; and α sums all the various loss mechanisms, such as free carrier absorption, light scattering by waveguide imperfections, and barrier material absorption, but not the fundamental absorption of the active layer, which is already included in the gain g definition. Equation 12.19 can be rewritten in the form:

$$\Gamma g_{th} = \alpha + \frac{1}{L} \ln \frac{1}{R} \qquad (12.20)$$

In GaAs/GaAlAs lasers, the first term is usually 10 cm^{-1} and the second 40 cm^{-1} for a 300 µ long laser and uncoated facet reflectivity of 0.3.

The threshold current is known once the gain versus injected current is determined. This can be readily done by using the carrier density as an input parameter. For that density one can calculate both the maximum gain and the required injection current from the known radiative and nonradiative recombination channels. As mentioned above, gain occurs only once significant inversion has been achieved.

Confinement of carriers in 2D structures as the main measure to obtain high carrier densities turned out to be a convenient way to reduce the required threshold current density. Several variants of QWs were developed to also meet the requirements of optical confinement.

- SQW: Each quantized well state introduces a 2D DOS equal to $m * \hbar^2$, while the onset of 3D states at the top of the well introduces a much larger DOS.

- Multiple quantum well: Each quantized state introduces a 2D DOS equal to $Nm * \hbar^2$, N being the number of wells.

- SCH laser: The intermediate-composition layers that surround the inner well introduce a large DOS, not as far apart from the ground state as in the DH, SQW, or multiple quantum well (MQW) cases.

- Graded-index SCH laser: The QW is surrounded by a region with steadily increasing refractive index. A ladder of quantum states in the graded region is added to the states of the inner well.

As discussed above, the SQW structure leads to $\Gamma \approx 4 \cdot 10^{-3}$, an unacceptable value, and one has to resort to the various structures shown in Fig. 12.8 [16, 17].

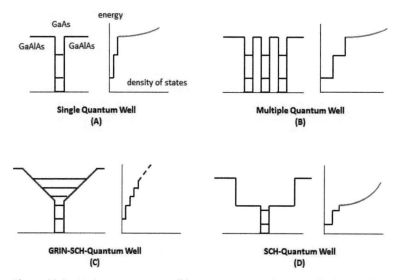

Figure 12.8 Various quantum well laser structures schematically depicted by their conduction band edge space variations (the left one in each set of figures) and their 2D density of states $Z(E)$ (right side).

12.5 Conduction Band Occupation in an Indirect Semiconductor

The indirect conduction band in an indirect semiconductor is occupied by more electrons than the direct one under equilibrium as a result of the Fermi–Dirac statistics. This general law of statistics can only be overcome with very fast transients, for example, injection of carriers at energies above the direct gap. The corresponding time frame is given by the lifetime of intervalley scattering, which is on the order of a few hundred femto seconds. The lifetime of the direct (Γ) – indirect (L) intervalley electron scattering in bulk Ge [18, 19] and

Ge/SiGe QWs is reported to be 230 fs and 185 fs [20], respectively. The high occupation of the direct transition was demonstrated in a femto second spectroscopy [21], with optical excitation leading to an optical gain (1300 cm^{-1}) comparable with direct semiconductors and 25 times larger than the steady-state gain. These results could pave the way to pulsed laser operation of indirect semiconductors with high efficiency.

12.5.1 Steady-State Occupation Model for Indirect Semiconductors

In the following parts we will assume Fermi–Dirac statistics for the conduction band and the valence band, with quasi-Fermi energy levels F_c and F_v in the injection regime.

For a simple model of the occupation statistics influence let us assume the same effective mass for the direct and indirect conduction bands. For low injection the Boltzmann approximation is valid. The ratio n_{dir}/n_{ind} of direct to indirect occupation is then simply given by Eq. 12.21:

$$\frac{n_{dir}}{n_{ind}} = \exp\left(-\frac{\Delta E}{k_B T}\right),\tag{12.21}$$

with ΔE the energy difference between the indirect and direct conduction bands. In Si the difference ΔE is very high (more than 2 eV) and the ratio n_{dir}/n_{ind} negligible (about 10^{-35}), so the upper level can be considered to be unoccupied. In Ge this energy difference shrinks to 0.136 eV or even lower for tensile-strained Ge. Then the upper level is also partly occupied and the ratio is small, for example, $5 \cdot 10^{-3}$ at room temperature and the simplifications made in Eq. 12.21. But even with this small ratio of below 1% direct transition dominates light emission because of the much higher radiative transition efficiency.

At medium and higher injection levels the ratio n_{dir}/n_{ind} increases above the fixed value (Eq. 12.21) of the Boltzmann approximation. This increase of the ratio n_{dir}/n_{ind} starts when the quasi-Fermi level F_c enters the indirect conduction band ($F_c - E_{cind} > 0$). It follows from simple mathematical properties of the modified Fermi–Dirac integral $F_{1/2}$, which describes the carrier density from folding of DOS with Fermi–Dirac statistics in bulk (3D) materials.

$$n = N_c F_{1/2}\left((F_c - E_c)/kT\right) \tag{12.22}$$

The effective DOS N_c is given by

$$N_c = \left(\frac{2\pi m^* kT}{h^2}\right)^{3/2} M_c, \tag{12.23}$$

with M_c as the number of equivalent minima (degeneracy of the band edge). The Fermi integral $F_{1/2}$ increase in the Boltzmann approximation regime as

$$F_{1/2}((F_c - E_c/kT)) = \exp((F_c - E_c)/kT), \tag{12.24}$$

where

$$(F_c - E_c)/kT < 0,$$

but for positive values of $(F_c - E_c)/kT$ the slope is smaller than for an exponential curve.

The modified Fermi–Dirac integral $F_{1/2}$ is normalized by a near unity prefactor ($2\sqrt{\pi} = 1.128$) to get exactly the exponential function for the Boltzmann approximation regime (Eq. 12.22).

The definition of the modified Fermi–Dirac integral of the order ½ is given by [22]

$$F_{\frac{1}{2}}(x) = \frac{2}{\sqrt{\pi}} \int_0^\infty \frac{u^{1/2}}{1 + \exp(u - x)} du. \tag{12.25}$$

(Please note that the Fermi–Dirac integral is sometimes used without the prefactor, for example, in the popular book in Ref. [23]. The prefactor has then to be added to the relation 12.22).

The deviation of $F_{\frac{1}{2}}(x)$ from $\exp(x)$ is clearly visible at positive values of the argument x as shown in Table 12.3, which lists the ratio R between exponential function and modified Fermi–Dirac integral.

$$R = \frac{\exp(x)}{F_{\frac{1}{2}}(x)} \tag{12.26}$$

The connection to statistics is obvious if one uses the substitution given in Eq. 12.27.

$$\chi = (F_c - E_c)/kT \tag{12.27}$$

The deviation from Boltzmann approximation ($R > 1$) increases strongly with the quasi-Fermi level entering the conduction band, with R already reaching a value of 16.8 at $(F_c - E_c) = 5$ kT (around 125 meV at room temperature).

Conduction Band Occupation in an Indirect Semiconductor | 287

Table 12.3 The ratio R is tabulated for the range from Boltzmann approximation ($x < 0$) to positive values of x

x	-4	-3	-2	-1	0	1	2	3	4	5
R	1.00	1.02	1.05	1.12	1.31	1.72	2.62	4.5	8.4	16.8

The same relation (Eq. 12.26) holds for the hole quasi-Fermi energy F_v in the valence band but with an inverted energy scale.

$$x = \frac{E_v - F_v}{kT} \tag{12.28}$$

For the simple model with the same effective direct/indirect masses an approximation [24] is given for quasi-Fermi energies above the indirect band edge but below the direct band edge.

$$\ln\left(\frac{n_{\text{dir}}}{n_{\text{ind}}}\right) = -\frac{\Delta E_C}{kT} + \left(\frac{3}{4}\sqrt{\pi}\right)^{2/3}\left(\frac{n_{\text{ind}}}{N_c}\right)^{2/3} - \ln\left(\frac{n_{\text{ind}}}{N_c}\right), \tag{12.29}$$

where

$$\left(\frac{n_{\text{ind}}}{N_c}\right) > 1 \text{ and } \left(\frac{n_{\text{dir}}}{N_c}\right) < 1.$$

The first term of Eq. 12.27 describes the Boltzmann approximation, and the second and third terms describe the influence of the Fermi integral $F_{1/2}$ on the occupation statistics of the indirect conduction band. The value of the ratio $\frac{n_{\text{dir}}}{n_{\text{ind}}}$ in Eq. 12.29 is plotted in Fig. 12.9 as a function of the relative indirect electron concentration $\left(\frac{n_{\text{ind}}}{N_c}\right)$ in a double logarithmic presentation.

This is done on a simplified (same effective mass direct and indirect electrons) hypothetical two-band system to demonstrate the influence of Fermi–Dirac statistics. The higher the electron density, the higher is the ratio of direct electrons to indirect electrons, which means that radiative recombination from direct states increases superlinearly with electron concentration. This simple outcome from general Fermi–Dirac statistics plays an important role in the design of indirect semiconductor lasers, which benefits from a larger ratio of direct-to-indirect transitions. This variable ratio of direct-to-indirect states is popularly described as the quasi-direct character of the highly n doped indirect semiconductor. However, the number of direct electrons is always lower than the indirect ones, although the ratio improves when the Fermi energy enters the conduction band.

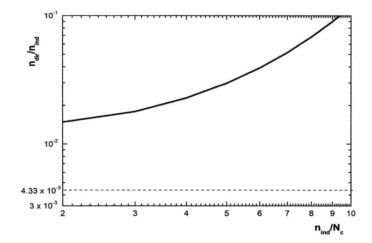

Figure 12.9 Presentation of the ratio $\left(\dfrac{n_{dir}}{n_{ind}}\right)$ versus relative electron concentration $\left(\dfrac{n_{ind}}{N_c}\right)$ in a double logarithmic plot. Assumed are same effective masses for direct and indirect electrons of a hypothetical indirect semiconductor with $\Delta E_c = 136$ meV.

12.5.2 Effective Density of State for Ge

The band edge of the band structure may be described by the parabolic dependence of the energy E on the wave number because a steady function $E(k)$ has no linear term at the extrema points. The effective mass is then given by

$$\frac{\partial^2 E}{\partial k^2} = \frac{\hbar^2}{m^*}. \tag{12.30}$$

An effective mass independent from the direction is typical for the direct conduction band state, which means a spherical constant energy surface in the $E(k_x, k_y, k_z)$ representation. For the indirect transition the $E(\vec{k})$ relation near the maximum energy is written as

$$E - E_{cind} = \frac{\hbar^2}{2}\left(\frac{(k_c - k_0)^2}{m_e} + \frac{k_t^2}{m_t}\right), \tag{12.31}$$

which may be presented by an ellipsoidal constant energy surface. The direction toward the minimum energy shows the longitudinal effective mass m_e. Both transversal directions have a different transverse effective mass m_t. The mean value for the effective DOS calculations is the DOS effective mass m^*_{DOS}, which is written in this chapter as m^* as we do not need any other mean value calculation (e.g., for conductivity)

$$m^* = (m_e^* m_t^2)^{3/2} \qquad (12.32)$$

In the valence band the extremum of the energy (maximum) is always direct at $k = 0$. Two sub-bands, heavy hole (hh) and light hole (lh), are degenerate at the BZ center. The band curvature $\dfrac{\partial^2 E}{\partial k^2}$ is not isotropic for the direct valence band but is given by a nonisotropic [25] energy wave number function $E(k)$

$$\frac{2m_0}{\hbar^2}(E_v - E) = Ak^2 \pm \left[B^2 k^4 + C^2 \left(k_x^2 k_y^2 + k_y^2 k_z^2 + k_z^2 k_x^2 \right) \right]^{1/2} \qquad (12.33)$$

The plus/minus sign in Eq. 12.33 is valid for the hh/lh sub-bands. The valence band parameters A, B, and C are given in Table 12.4 [25].

Table 12.4 Valence band parameters A, B, and C for Ge (in comparison with Si) (compiled from data given in Ref. [25])

	A	B	C	SOED
Ge	−13.3	−8.6	12.8	290 meV
Si	−4.3	−0.6	4.9	44 meV

Note: Spin-orbit energy difference (SOED) for the split-off band is added to show the rather high energy difference for Ge.

Along the cubic axes the effective masses are $\dfrac{m_0}{A+B}$ and $\dfrac{m_0}{A-B}$ for the light holes and the heavy holes, respectively. A third sub-band, the spin-orbit split-off band, need not usually be considered for occupation statistics in Ge because of the rather large energy separation SOED of 290 meV. The effective DOSs for the different sub-bands were calculated with the use of Eq. 12.23. The number of equivalent minima M_c for the indirect band is 4, and for all other sub-bands $M = 1$ was chosen. The values of the effective masses differ slightly in literature because of experimental uncertainty and deviations from parabolicity of the band structure. We have chosen here the values selected in Ref. [23].

The effective masses and the calculated effective DOS for the four sub-bands (in the conduction band indirect and direct sub-bands with subscripts ind and dir, respectively. In valence band heavy hole and light hole sub-bands with subscripts hh and lh, respectively) are summarized in the Table 12.5.

Table 12.5 The effective masses m^* of the Ge conduction and valence sub-bands in units of the free electron mass m_o

Band	Conduction		Valence	
Sub-band	ind	dir	hh	lh
m^*/m_o	0.22	0.038	0.28	0.044
N_c/N_v (units 10^{18} cm^3)	10.75	0.185	3.7	0.23

Note: The subscripts ind, dir, hh, and lh refer to the indirect and direct conduction sub-bands and heavy hole and light hole valence sub-bands. The corresponding effective densities of states are calculated from Eq. 12.23.

The effective DOS of a nominal band with free electron mass m_o is given as 2.5×10^{19}/cm^3. The effective DOS of the different Ge sub-bands varies considerably from the 10^{17}/cm^3 range (dir, lh) to the 10^{18}/cm^3 range (hh) and the onset of the 10^{19}/cm^3 range (ind). The effective density of the band is the weighted sum of the sub-band densities within the Boltzmann approximation.

$$N_c = N_{cind} + N_{cdir} * \exp\left(\frac{-\Delta E_c}{kT}\right) \qquad (12.34)$$

The much higher density N_{cind} of the indirect band and the energy distance E_c of the direct band allow the contribution of the direct band to be rather low (about 10^{-4}) so that $N_c = N_{cind}$. This holds true even up to the indirect/direct crossover $\Delta E_c = 0$, which can be obtained with very high tensile strain or Sn alloying.

For the valence band of an unstrained group IV semiconductor we obtain

$$N_v = N_{vhh} + N_{vlh}. \qquad (12.35)$$

The valence band degeneracy at $\bar{k} = 0$ is lifted by an applied strain. Under tensile strain the lh sub-band is the higher lying. Then we obtain with an energy difference δE_v between both sub-bands hh and lh

$$N_v = N_{vlh} + N_{vhh}*\exp(-\delta E_v). \qquad (12.36)$$

At the valence band we see a much lower density from the lh sub-bands (0.23 versus $3.7 \cdot 10^{-18}/cm^3$ for the hh band) so that the density $N_v = 4 \cdot 10^{-18}/cm^3$ is mainly determined by the hh sub-band. Tensile strain reduces the density N_v by lowering the weight of the density N_{vhh} (Eq. 12.36). However, an energy difference of the sub-bands of $3k_BT$ (75 meV at room temperature) is necessary for a reduction of N_v toward the N_{vlh}.

A definition of a common DOS for nondegenerate sub-bands with an energy difference ΔE is not possible outside the Boltzmann approximation. Then the carrier density n (or p) has to be written as the sum of both sub-bands.

$$n = N_{c1}F_{1/2}(x) + N_{c2}F_{1/2}(x - \Delta x), \qquad (12.37)$$

with the abbreviations $x = (F_c - E_c)/kT$, $\Delta x = \Delta E/kT$.

The following Table 12.6 gives a comparison of the Fermi–Dirac integral F (order ½) with the Boltzmann approximation.

Table 12.6 Fermi–Dirac integral $F_{1/2}$ in comparison with the exponential functions $\exp(x)$

$x = (F_c - E_c)/kT$	$F_{1/2}(x)$	$\mathbf{Exp}(x)$
-4	$1.8 \cdot 10^{-2}$	$1.8 \cdot 10^{-2}$
-3	$4.9 \cdot 10^{-2}$	$5 \cdot 10^{-2}$
-2	$12.9 \cdot 10^{-2}$	$13.5 \cdot 10^{-2}$
-1	$32.8 \cdot 10^{-2}$	$36.8 \cdot 10^{-2}$
0	$76.5 \cdot 10^{-2}$	1
1	1.58	2.71
2	2.82	7.38
3	4.49	20.1
4	6.51	54.6
5	8.84	148.4

Many rather accurate approximations of the Fermi–Direct integral are available. A rather easy intrinsic form is suggested by Kroemer [26] for arguments $x > 1$.

$$x = \ln(F_{1/2}(x)) + 0.35\, F_{1/2}(x) \qquad (12.38)$$

The rather simple Kroemer approximation (Eq. 12.38) is accurate enough to be used together with the effective mass approximation because both are very good for Fermi energies near the band edges and both lose some accuracy at higher Fermi energies but are good to predict systematic trends. For higher numerical accuracies the full band structure has to be considered.

12.6 Influence of Strain and Sn Alloying

A necessary condition for an active material in a semiconductor laser is the population inversion that starts when the quasi-Fermi energies F_C and F_V enter the conduction band and the valence band, respectively. Then the gain g (Eq. 12.5) switches from a negative value (stimulated absorption dominates) to a positive value (stimulated emission dominates). In a real-world laser the gain at threshold g_{th} (Eq. 12.20) has to be larger than zero in order to counterbalance modal losses (described by the confinement factor Γ) and extinction losses (described by the coefficient α'). Modal losses and extinction losses depend on the dedicated device structure, so that we follow in the fundamental material discussion the necessary condition $g = 0$. Next, we consider an n-doped active device in which a nonequilibrium carrier density is injected. In the active material region the total carrier density is then given as

$$\text{Electrons} \quad n = N_d + \Delta n$$

and

$$\text{Holes} \quad p = \Delta p.$$

The higher electron charge $-qn$ is counterbalanced by the lower charge $+ qN_D$ and the hole charge $+ q\Delta p$. For the calculation of the Fermi energy levels F_C and F_V in a two-sub-band system Eqs. 12.36 and 12.37 may be used. We obtain for the Fermi energy touching the direct conduction band the relation:

$$N_d + \Delta n = N_{ind} F_{1/2}(\Delta X_c) + N_{dir} F_{1/2}(0) \qquad (12.39)$$

For the strain-split valence band we obtain Eqs. 12.40 and 12.41, respectively.

$$\Delta p = N_{\text{vlh}} F_{1/2}(0) + N_{\text{vhh}} F_{1/2}(\delta X_{\text{v}}) \qquad (12.40)$$

Equation 12.40 is valid for tensile strain when the lh band is on top.

$$\Delta p = N_{\text{vhh}} F_{1/2}(0) + N_{\text{vlh}} F_{1/2}(\delta X_{\text{v}}) \qquad (12.41)$$

Equation 12.41 is valid for compressive strain when the hh band is on top.

For clarity of presentation we use for the band parameters N_{V} and N_{C} the same values as for unstrained Ge. These values are given in Table 12.5. Let us explain the treatment for strained Ge where the normalized quantities are defined as

$$\Delta X_{\text{c}} = \frac{E_{\text{cind}} - E_{\text{cdir}}}{k_{\text{B}} T} \qquad (12.42)$$

and

$$\delta X_{\text{v}} = \frac{E_{\text{Vlh}} - E_{\text{Vhh}}}{k_{\text{B}} T}. \qquad (12.43)$$

At room temperature unstrained Ge has

$$\Delta X_{\text{c}} = 136 \text{ meV}/25 \text{ meV} = 5.44$$

and

$$\Delta X_{\text{v}} = 0.$$

Using the values of Tables 12.5 and 12.6 one obtains $p = \Delta p = 3 \times 10^{18}/\text{cm}^3$ and $n = N_{\text{d}} + \Delta n = 1 \times 10^{20}/\text{cm}^3$. The necessary condition $g = 0$ will be obtained at room temperature in unstrained Ge with a very high n-doping $(10^{20}/\text{cm}^3)$ and a moderate injection density level of $\Delta p = \Delta n = 3 \times 10^{18}/\text{cm}^3$.

The very high n-doping is problematic both for technological reasons as well as for increasing the free carrier absorption and reducing the carrier lifetime by Auger recombination. The need for very high n-doping may be reduced effectively by smaller values of the direct-indirect bandgap energy ΔE_{c}. This energy difference shrinks with larger lattice spacing, which may be obtained by tensile strain or alloying with a larger atom like Sn (or maybe with the even larger Pb atom in the future).

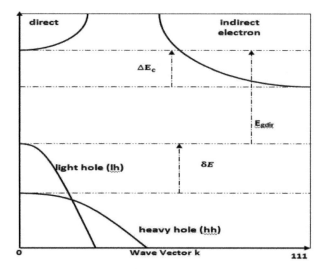

Figure 12.10 Band structure of tensile-strained Ge (biaxial stress). The essential quantities for laser operation are the energy difference ΔE_c between indirect and direct conduction bands, the direct bandgap E_{gdir}, and the energy splitting δE between heavy hole and light hole band. The bandgap E_{gdir} is not shown to scale.

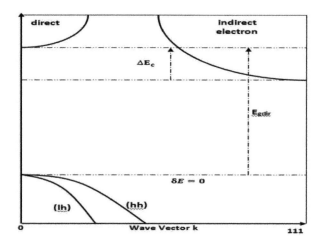

Figure 12.11 Band structure of unstrained GeSn.

The much higher effective DOS of the indirect band (roughly 50 times higher than that of the direct one, see Table 12.5) dominates the carrier density not only in the indirect semiconductor region

($\Delta E_c > 0$) but also at the crossover ($\Delta E_c = 0$) till deep into a direct semiconductor ($\Delta E_c = -4k_BT$) behavior. At crossover the necessary electron density is about $1.1 \cdot 10^{19}/cm^3$, and it drops to about $3 \times 10^{17}/cm^3$ at a direct semiconductor with $\Delta E_c = -100$ meV. The valence band splitting under the action of strain has to be considered in order to calculate the optimum doping and the injected carrier density.

The most important technical situations are given by a Ge (or GeSn) under tensile strain (Fig. 12.10), by unstrained GeSn (Fig. 12.11), and by compressive strained GeSn (Fig. 12.12). As mentioned above the tensile strain reduces ΔE_c and it splits the valence band, with the lh band on top (Fig. 12.10) and the hh band below (δE_v at the BZ center at $\bar{k} = 0$). The necessary tensile strain for a crossover to a direct semiconductor is high, about 2%–5%, but in detail depending on the strain direction (uniaxial, biaxial) and the assumed deformation potential parameters. Unstrained GeSn enhances the lattice constant, reduces the bandgap E_{gdir} and the difference ΔE_c, and does not lift the degeneracy of lh and hh valence bands (Fig. 12.11). The crossover to a direct semiconductor takes place at about 8% Sn content. Technical handling and strain management in GeSn are limited by its low temperature fabrication window. Therefore, often devices with GeSn in compressive strain are fabricated. These device structures need higher Sn content for ΔE_c reduction and the hh band is on top of the valence band structure (Fig. 12.12).

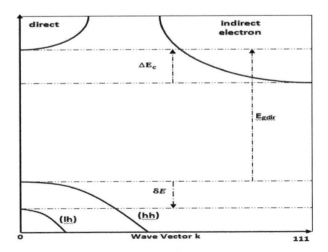

Figure 12.12 Band structure of compressive stained GeSn.

12.6.1 Injected Carrier Densities and Layer Doping

From the requested quasi-Fermi energies F_c and F_v one can undoubtedly derive the minimum carrier levels n and p for gain $g = 0$. From these we can calculate the recommended doping level for the condition of minimum injected carrier densities Δn and Δp.

Key parameters for the carrier levels are the energy difference ΔE_c between the indirect conduction band edge and the direct one (Fig. 12.10) and the energy difference δE_v between the hh and the lh bands. The electron concentration is determined by the parameter ΔE_c and the hole concentration p by the parameter δE_v. The energy difference ΔE_c is a function of Sn and strain, whereas the lh-hh energy difference δE_v is a function of strain alone.

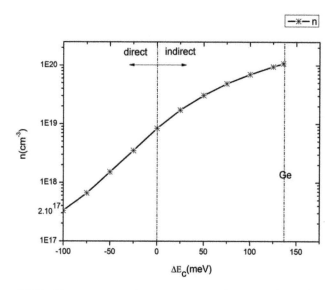

Figure 12.13 The necessary electron concentration n to obtain occupation inversion with gain $g = 0$, given as a function of the direct/indirect gap ΔE_c.

Figure 12.13 shows the dependence of the electron concentration n on the energy difference ΔE_c. The numerical values are obtained with the material parameters of Ge, but the general trend is independent of that. The necessary concentration of electrons is very high (about $10^{20}/cm^3$) for unstrained Ge ($\Delta E_c = 136$ meV) and drops down to yet higher values (about $10^{19}/cm^3$) at the direct/

indirect band crossover ($\Delta E_c = 0$), and it continues to decrease to direct bandgap regime ($\Delta E_c < 0$) with a low saturation level (about $10^{17}/cm^3$) for a strong direct semiconductor behavior ($\Delta E_c < -150$ meV).

Figure 12.14 shows the dependence of the hole concentration p on the energy difference δE_v. Note, we use positive values of δE_v for tensile strain (lh sub-band above hh sub-band at wave vector $\bar{k} = 0$) and negative values for compressive strain. The hole concentration for compressive strain and unstrained material is moderately high (about $3 \cdot 10^{18}/cm^3$), and it drops sharply down to a lower level with tensile strain reaching a saturation (about $2 \cdot 10^{17}/cm^3$) at strong tensile strain ($\delta E_v > 125$ meV).

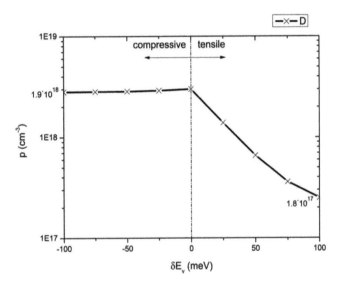

Figure 12.14 The necessary hole concentration p (for $g = 0$) given as a function of the strain related lh-hh energy gap δE_v.

What are the consequences of the data in Figs. 12.13 and 12.14 for the optimum doping and carrier injection levels? The electron concentration in the indirect semiconductor regime ($\Delta E_c > 0$) is always considerably higher than the hole concentration. This means the carrier injection level is given by the requested p concentration whereas the much higher electron concentration is mainly provided by the higher n-doping N_D, which ranges from $10^{20}/cm^3$ for unstrained Ge to about $5 \cdot 10^{18}/cm^3$ for the indirect/direct crossover material (either obtained by Sn alloying or a tensile strain or a combination of both). The minimum injection carrier density $\Delta n = \Delta p$ is given by the requested hole density, and it depends on the strain splitting of both hole sub-bands. The injected carrier density ranges from about $3 \cdot 10^{18}/\text{cm}^3$ for compressive and unstrained material to $2 \cdot 10^{17}/\text{cm}^3$ for highly tensile strained material. Therefore, we see two beneficial effects of the tensile strain within the indirect bandgap regime, the n-doping level is lower compared with the unstrained material because of a lower ΔE_c and the injected carrier density is lower because the minimum $\Delta n = \Delta p = p$ decreases by a factor of 15 with high tensile strain. The situation seems to be more complex in the direct semiconductor regime because the necessary n and p values cover similar ranges from several $10^{18}/\text{cm}^3$ to $10^{17}/\text{cm}^3$ so that theoretically a p-doping could be necessary for compressive strained and unstrained material with $\Delta E_c < -25$ meV. In practice a compressive strained Sn needs very high Sn contents to obtain a direct semiconductor state, which is technically complex and not very favorable for an Si-based device. For unstrained GeSn, we see an undoped GeSn as optimum (n=p) for a slight direct GeSn (around a difference = -25 meV) and we recommend a slight p-doping in the 10^{17}–$10^{18}/\text{cm}^3$ range for more direct GeSn alloys. Details of the influence of Sn and strain ε on the band parameters ΔE_c and δE_v are given in Chapter 11. Here we need only rough values to sketch the general trends. For this reason we simplify the Sn influence by a linear function with a direct/indirect crossover at about 8% Sn. Then with about 1.5% Sn increase the direct band approaches by 25 meV, for example, with 6% Sn it is an indirect semiconductor ($\Delta E_c \approx$ +35 meV) and with 10.5% it is a direct one ($\Delta E_c \approx -40$ meV). Tensile strain reduces the energy gap ΔE_c between direct band and indirect one. For a first assessment, a linear dependence on strain may be assumed. The valence band splitting δE_v can only be treated linearly at small strain values because the hh sub-band moves linearly with strain but the lh follows a square root law [27]. Around 5% uniaxial strain and around 2% biaxial strain are needed to get a crossover of direct and indirect conduction bands [28].

Tensile-strained GeSn (containing about 10% Sn) with low doping levels would fulfill demands for low injection and low carrier concentration because the sub-bands defining the band properties— the direct conduction band and the lh valence band—both need low concentrations of carriers for population inversion.

12.6.2 Internal Quantum Efficiency and Threshold Current

The electrically stimulated semiconductor laser is the most relevant laser type for applications. The crucial parameters for the electrical operation are the threshold current I_{th} for the onset of laser emission. This threshold requires a positive net gain G ($G > 0$). In the section before, we have seen that the gain is correlated with the necessary carrier densities in the active medium. In this part, the connection between injection currents and carrier densities is given for idealized carrier confinement conditions. On this basis the internal quantum efficiency (IQE) may be treated as a competition between radiative and nonradiative recombination processes. The simplest electrical laser structure consists of a p/n junction diode. Here, the injected current induces a deviation from the equilibrium that reaches more than a diffusion length L_d apart from the junction. Carrier confinement is obtained very effectively by a heterostructure [12]. For a Ge-based laser, larger bandgap materials, such as Si [29] and GeSiSn [30], may be utilized as barrier materials, for example, for a Si/Ge/Si double heterobarrier structure. The injection efficiency for such a heterostructure carrier confinement structure is ideally assumed to be perfect, 100% [29], which means all the injected carriers are provided within the active layer for recombination, radiative and nonradiative. The IQE is defined under these idealizing conditions as the ratio between radiative recombination rate U_{rad} and total recombination rate, which is written as the sum of U_{rad} and nonradiative recombination rate U_{nrad}.

$$IQE = \frac{U_{rad}}{U_{rad} + U_{nrad}} \tag{12.44}$$

In unstrained Ge, the radiative recombination rate U_{rad} is already dominated by the direct transition between the Γ minimum in the conduction band and the valence band. The dominance of direct radiative recombination in comparison with the indirect one will be even larger with reduced ΔE_c in GeSn and strained Ge. Earlier observations of comparable indirect and direct emissions from substrate or thick layers were explained by self-absorption of the more energetic direct emission [31]. The radiative recombination rate is given by

$$\left(\frac{n_{\text{dir}}}{n}\right)pU_{\text{rad}} = R_{\text{dir}}.n_{\text{dir}} = R_{\text{dir}}n, \tag{12.45}$$

with R_{dir} the recombination rate for direct recombination. It is proportional to the concentration of the direct electrons n_{dir} and of holes. The competing nonradiative recombination is dominated by the acceleration of electron-hole annihilation at midgap recombination centers (Shockley–Read–Hall [SRH] effect) and by Auger recombination. In an n-type semiconductor both these effects contribute to U as

$$U_{\text{nrad}} = \frac{p}{\tau_{\text{SRH}}} + C_n n^2 p. \tag{12.46}$$

The recombination at the mid-bandgap center is described by the minority carrier life time τ_{SRH}, which is smaller in a semiconductor with defects. Defects may be dislocations, vacancies or interstitials, and metallic contaminations. These defects can be easily produced by a lattice mismatch, nonequilibrium growth conditions, and improper device processing. Especially the Saraswat group [32] has addressed the importance of high-quality active layers with a long minority carrier lifetime for Ge-based lasers with good IQE. At very high carrier concentrations the Auger effect dominates because of its quadratic dependence on the electron concentration (in n-material). In an Auger process two electrons with opposite momentums meet a hole with zero momentum and they can therefore recombine without the help of a phonon. The defect-related minority carrier lifetime and the Auger recombination may be combined to a concentration-dependent total nonradiative lifetime for which the following relation holds:

$$\frac{1}{\tau_{\text{nrad}}} = \frac{1}{\tau_{\text{SRH}}} + C_n n^2 \tag{12.47}$$

Figure 12.15 shows the resulting inverse nonradiative lifetime $\left(\frac{1}{\tau_{\text{nrad}}}\right)$ as a function of the electron concentration in a double logarithmic presentation with the defect-related lifetime as a parameter. The Auger recombination coefficient at room temperature is chosen as 10^{-30} cm^6/s [33]. The defect-related minority carrier lifetime varies between below 1 ns for a material with high dislocations and point defect densities (named "bad material" here) and much more than 100 ns (good material) for materials with low defect densities. The Auger process dominates the recombination at very high electron concentrations (10^{20}/cm^3). Then even a good material has high nonradiative recombination rates. The advantages of good materials (with respect to lifetime) are only visible at lower electron levels (10^{19}/cm^3), which are obtained in laser structures with highly strained Ge or GeSn with above 5% Sn.

The IQE is computed by combining Eqs. 12.44–12.46.

$$\text{IQE} = \frac{R_{\text{dir}} n(n_{\text{dir}}/n)}{1/\tau_{\text{SRH}} + C_n n^2}. \tag{12.48}$$

This relation is also valid for the LED regime if the perfect carrier confinement is fulfilled.

A high defect recombination rate (low SRH lifetime τ_{SRH}) reduces the quantum efficiency of emission, which is intuitively clear as a nonradiative channel is opened. But a low lifetime also influences the threshold current I_{th}, which is necessary to establish the carrier levels requested for a positive net gain G.

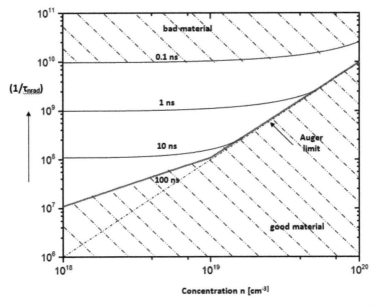

Figure 12.15 Dependence of the inverse of nonradiative lifetime ($1/\tau_{\text{nrad}}$) from the electron concentration n for different defect-related lifetimes τ_{SRH}.

Let us assume some conditions as for quantum efficiency calculations, which means perfect carrier confinement and n-doping of the active layer. The continuous loss of carriers by the recombination has to be counterbalanced by the injected carrier of the electrical current.

$$I = J \cdot A = qV(U_{\text{rad}} + U_{\text{nrad}}), \qquad (12.49)$$

where I is for current, J for current density, A for cross section, V for volume, and U for recombination rate (per volume).

With $V = A \cdot d$ (d is the thickness of the confinement region) we use Eqs. 12.45, 12.46, and 12.49 to obtain

$$J_{\text{th}} = q^* p_{\text{th}} \left(R_{\text{dir}} n_{\text{dir}} + 1 / \tau_{\text{SRH}} + C_n n_{\text{th}}^2 \right) \qquad (12.50)$$

if $p_{\text{th}} = \Delta n = \Delta p$ and $n_{\text{th}} = N_D + p_{\text{th}}$ are set to the values needed for the net gain $G = 0$. Please remember that we have given in the last section a recipe for the optimum choices of doping and carrier densities to get the gain $g = 0$. The injected carrier densities required to get the net gain $G > 0$ are higher because the modal confinement ($\Gamma \leq 1$) and the extinction coefficient α' request a positive gain g_{th} to obtain a net gain $G = 0$ (see Eq. 12.20). A general rule for the choice of Γ and α' is not available because these parameters depend on the device design. The extinction coefficient α' contains device-specific components such as absorption from contact layers and surface scattering, but also an unavoidable component from free carrier absorption (the fundamental spontaneous absorption of the active layers is contained already in the gain definition and must not be added here).

The strong increase of the nonradiative recombination at high doping and carrier injection levels causes very high threshold currents I_{th}. For example, a threshold current density of 500 kA/cm^2 is necessary for a doping of 10^{20}/cm^3 and a carrier injection of 10^{19}/cm^3. The threshold current density shrinks to 5 kA/cm^2 for a doping of 10^{19}/cm^3 and a carrier injection of $5 \cdot 10^{18}$/cm^3. The Auger limit restricts continuous wave (CW) operation of Ge-based laser to strain and alloy conditions, which reduces the direct/indirect gap ΔE_c to below 100 meV and increases the lh-hh valence band splitting to above 0 meV.

12.7 Experiments

Direct bandgap semiconductors mostly from III/V compounds dominate the market for visible and infrared lasers of compact size and moderate power output. They use direct bandgap emission very efficiently. The low radiative emission rate of the indirect semiconductor silicon prohibited the use of the bulk silicon as an active laser medium. The silicon-based research community put a lot of effort to overcome the low radiative emission rate by different types of nanostructuring, by introducing defect levels, and by doping with color centers (rare earth elements). All this effort brought rich physical understanding and an enhancement of radiative emission but failed to push silicon as a laser material. The general opinion that an indirect semiconductor is useless for laser emission seemed to be confirmed. However, the situation with Ge is quantitatively very different from the viewpoint of the energy distance ΔE_c between the lower indirect band and the higher lying direct band. This energy difference ΔE_c is very high for Si (more than 2 eV, which is more than 80 kT at room temperature) but moderately small (0.0136 eV, which is about 5 kT at room temperature) for Ge. There were some rumors in the Western world that a Russian researcher had achieved gains in n-doped Ge but made no laser demonstration. Probably, the first group that spent continuous theoretical and experimental efforts for a Ge laser was the Kimerling group [34, 35] at MIT, USA. They calculated the gain spectrum g (hv) of n-doped Ge for different injection carrier levels $\Delta n = \Delta p$ and they found rather high gain values for a population inversion factor $(f_c - f_v) = 1$. Their approach is based on the inherent connection between stimulated emission and absorption (Eq. 12.5) and on the joint density of states model (Eq. 12.13) with a reduced carrier mass m_r (Eq. 12.2). The population inversion factor $(f_c - f_v)$ connects stimulated emission with the absorption coefficient

$$g(hv) = \alpha_{\text{dir}}(f_c - f_v) \qquad (12.51)$$

with α_{dir} as the absorption coefficient connected to the direct bandgap transition. That is easily demonstrated by switching the population inversion factor from equilibrium $(f_c - f_v) = -1$ to lasing conditions $1 > (f_c - f_v) > 0$. At equilibrium the gain g is $-\alpha$ (loss). This condition (Eq. 12.51) allows the calculation of the gain spectrum

304 | *Laser*

from the absorption coefficient and the quasi-Fermi levels F_c and F_v. The absorption coefficient α_{dir} may be acquired by a transmission experiment or from the known energy relation of direct absorption processes. Note that near the direct bandgap three absorption processes take place: the searched direct absorption, the indirect band absorption, and the free carrier absorption. The direct band-to-band absorption can be theoretically calculated by solving the electron-photon scattering in a crystalline potential [36, 37].

$$\alpha(hv) = \frac{q^2 hc\mu_0}{2m_0^2} \frac{P_{cv}}{n_r} \frac{1}{g(hv)} Z_j, \qquad (12.52)$$

where q, h, c, μ_0, m_0, and n_r have the usual meaning (electron charge, Planck's constant, speed of light, permittivity coefficient, free electron mass, and refractive index, respectively) and ρ_{cv} and Z_j are the transmission matrix element and joint DOS, respectively.

The joint DOS of Z_j is given within the effective mass model as

$$Z_j = 2\pi \left(\frac{2m_r}{h^2} \right)^{3/2} \left(hv - E_{g\Gamma lh} \right)^{1/2}, \qquad (12.53)$$

with the reduced mass m_r (Eq. 12.12). The absorption near the direct band edge may be simplified described by

$$\alpha(hv) = \frac{A \left(hv - E_{g\Gamma lh} \right)^{1/2}}{hv} \qquad (12.54)$$

if we combine all slowly varying properties into a constant A. The constant A for Ge is about

$$2.10^4 \text{ eV}^{1/2}/\text{cm}$$

The more experienced reader may notice that we did not discuss the influence of excitons on the emission/absorption spectrum. The influence of excitons is treated in textbooks on optical properties [36] and in a recent publication of the Menendez group [38].

We continue with the simplified treatment of the band-to-band transition.

The absorption edge of tensile-strained Ge moves toward lower energy because of the shrinkage of the direct bandgap. Since the lh or hh band separate under strain, two optical transitions make up the total absorption. Then, the absorption of tensile-strained Ge is

$$\alpha(h\nu) = \frac{A}{h\vartheta}\left[k_1\left(h\nu - E_{\text{grlh}}\right)^{1/2} + k_2\left(h\nu - E_{\text{grhh}}\right)^{1/2}\right], \quad (12.55)$$

where k_1 and k_2 are coefficients attributed to the contributions from different reduced masses.

$$k_1 = \frac{m_{\text{rlh}}^{3/2}}{(m_{\text{rlh}}^{3/2} + m_{\text{rhh}}^{3/2})}$$

$$k_2 = \frac{m_{\text{rhh}}^{3/2}}{(m_{\text{rlh}}^{3/2} + m_{\text{rhh}}^{3/2})}$$

$$k_1 + k_2 = 1$$

For Ge the normalized coefficients k_1 and k_2 are given as 0.68 and 0.32, respectively.

The calculated gain spectrum of an n-Ge ($N_D = 7 \times 10^{19}/\text{cm}^3$) under slight tensile strain (0.25%) is shown in Fig. 12.16. High n-doping of unstrained or slightly strained Ge is necessary to get the electron quasi-Fermi level toward the direct band edge.

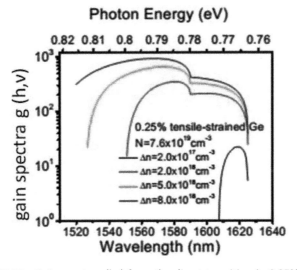

Figure 12.16 Gain spectra g(hv) from the direct transition in 0.25% tensile-strained n⁺-Ge with $N = 7.6 \times 10^{19}$ cm^{-3} at different injected carrier densities. Reprinted with permission from Ref. [35] © The Optical Society.

The injected carrier density is necessary to push the hole quasi-Fermi level toward the valence band. A positive gain is obtained from an injection level of $5 \cdot 10^{17}/\text{cm}^3$ ($\Delta n = \Delta p$) but the gain is small ($50/\text{cm}^2$) and restricted to a small band above the band edge (which is 0.7 eV for the assumed strain level). With higher injection levels the gain increases to about $1000/\text{cm}^3$ for $\Delta n = \Delta p = 8 \cdot 10^{18}/\text{cm}^3$. Clearly visible in the gain spectrum is the overlay of the separated lh and hh contributions where the quasi-Fermi level touches the bottom of the hh sub-band. Optical gain in a material under injection does not necessarily lead to a laser emission that requires the optical gain to overcome all kinds of optical losses. An unavoidable kind of loss is caused by the free carrier absorption within a band. The free carrier absorption increases with wavelength and becomes significant at higher carrier densities. An optimistic assessment of the lasing potential assumes that the optical confinement is perfect ($\Gamma = 1$) and the losses are caused by the free carrier absorption. Then the net gain G reduces simply to

$$\Gamma g - \alpha' = g - \alpha_{\text{fc}}, \tag{12.57}$$

with Γ as the optical confinement factor, g as the gain (a sum of all extinction losses except the fundamental absorption), α_{fc} as the free carrier absorption coefficient. The free carrier absorption scales with the carrier concentration and with the square of the wavelength (Drude model). A fit of experimental data [35] at room temperature results in

$$\alpha_{\text{fc}}(\lambda) = -3.4 \cdot 10^{-25} n\lambda^{2.25} - 3.2 \cdot 10^{-25} p\lambda^{2.43}, \tag{12.58}$$

where α_{fc} is measured in units of cm^{-1}, n and p are measured in units of cm^{-3}, and λ is measured in units of nm. The free hole absorption is higher than the free electron absorption, roughly 4 times higher at a wavelength of 2 µm. At this wavelength an optical gain of $1000/\text{cm}$ exceeds the free carrier absorption up to $1.11 \cdot 10^{20}$ electrons but only up to $3 \cdot 10^{19}$ holes. The conditions for Ge lasing with many more electrons than holes benefit from the lower free electron absorption.

The technical realization of such a laser needs good optical confinement, low losses from surface scattering, and good electrical confinement for threshold current reduction. Figure 12.17 shows a device suggestion [34] from the Kimerling group.

Experiments | 307

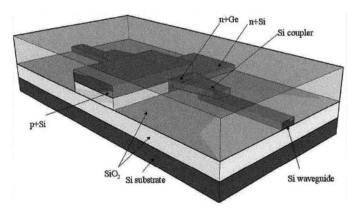

Figure 12.17 Proposed schematic of the integration of a Ge laser with a Si waveguide. Reprinted from Ref. [34] with permission from the Massachusetts Institute of Technology ("MIT").

A silicon-on-insulator (SOI) substrate gives Si waveguides with good optical confinement. The waveguide is connected to the laser device by a tapered Si coupler. The laser device itself is composed of an n^+-Si/n-Ge/p^+-Si double heterostructure (DH) with good carrier confinement.

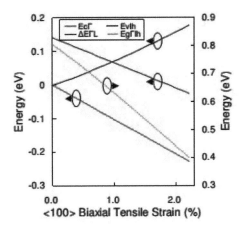

Figure 12.18 Strain dependency of Ge direct bandgap $E_{g\Gamma lh}$ (energy scale on the right). The energy scale on the left is valid for the conductance band Γ to L valley separation $\delta E = E_{\Gamma L}$, the lh valence band edge E_{Vlh}, and the direct conductance band edge $E_{C\Gamma}$, $E_{C\Gamma G}$ and E_{Vlh} are plotted with respect to their unstrained positions. Reprinted with permission from Ref. [39] © The Optical Society.

Liu et al. [35] calculated with the above-given data the net gain G for a bulk n$^+$-Ge (N_D = 1 × 10^{20}/cm^3) under tensile strain (ε = 0.25%). The bulk Ge needs higher doping (10^{20}/cm^3) and higher injection (Δn = 1 × 10^{19}) compared to slightly strained Ge with N_D = 7 × 10^{19}/cm^3 and Δn = 3.5·10^{18}/cm^3 in order to get a positive net gain G. Wada et al. [39] suggested micromechanical methods to increase the tensile strain level. Figure 12.18 shows the bandgap parameter as a function of biaxial strain.

12.7.1 Experimental Verification of Laser Light Emission from Group IV Materials

In Table 12.1, three different main configurations for laser in Si-based photonic systems are defined:

- Monolithic integration with Si photonics
- Hybrid mounting of group III/V laser on Si photonic systems
- Light coupling from an external laser source

All these system configurations have their own pros and cons, but their maturity differs greatly. Light coupling from an external source is now a de facto standard that will be partly replaced in the future by hybrid and monolithic approaches. Hybrid approaches use the well-developed III/IV lasers. Si photonics–based efforts are mainly concerned with mounting or bonding techniques for evanescent coupling of the light wave from the top-lying thinned laser to the Si-based waveguide below [9, 10, 40–44]. The details of the laser structure, the thinning procedure, the bonding, and the III/V laser for Si waveguide coupling schemes are sophisticated as, for example, explained in Ref. [42]. There, a quaternary InAlGaAs MQW laser is bonded to an SOI waveguide with a 500 nm device layer on a 1 µm thick buried oxide. The 1.5 µm wide silicon waveguide was defined by etching, as also the first-order grating on the Si waveguide, to create the resonator for the distributed feedback (DFB) laser. A DFB laser diode is a good candidate for the light source in wavelength division multiplexing systems. Single-frequency emission, short cavity length (100–200 µm), moderate threshold current density (1.1 KA/cm^2), and high-speed direct modulation (12.5 Gb/s) were reported. This work is used as a reference for comparing the efforts with monolithically integrated laser variants. The monolithic laser variants target in two directions with either III/V epitaxy on the substrate or group IV materials with strained Ge and GeSn. III/V epitaxy is a good choice for the laser part [45, 46] because of the excellent understanding of laser action and good optical gain, but the acceptance and frequent use of III/V materials with Si technology remain an open question. Electrically stimulated group IV lasers on Si could fill the gap in monolithically integrated photonic systems despite their imperfect performance in realized lasers [47]. Promising work for Ge-on-Si lasers was done by the MIT group of Kimerling et al. They demonstrated first [11] in 2010 an optical stimulation and two years later (2012) an electrical stimulation of a Ge laser on Si [12]. As the active laser material they used n-doped Ge with a slight tensile strain ($\varepsilon = 0.2\%$). The slight tensile strain was simply produced by annealing the Ge-on-Si laser to temperatures above 750°C. The thermal expansion mismatch causes the slight strain in the before-strain-relaxed Ge. This experimental proof was important not only for the technical development of on-chip lasers but also for clarification of a basic discussion on the possible net gain in Ge. Carroll et al. [48] doubted that the gain in Ge can overcome the fundamental losses, and they concluded from pump probe experiments that the losses are much higher than the usually assumed free carrier absorption. Koerner et al. [49] confirmed the experimental finding of the MIT group by even driving an unstrained Ge laser to the then very high (500 kA/cm^2) threshold current density. High tensile strain and alloying with Sn should reduce the threshold current density. Wirths et al. [50] demonstrated the fundamental use of GeSn as the active gain material with optical excitation at low temperatures. Here, progress in the synthesis of metastable GeSn alloys by epitaxy should deliver room temperature lasers with low threshold current and shift of the emission wavelength from the telecommunication wavelength 1.55 μm toward the midinfrared (2–3 μm) wavelength. Table 12.7 gives a comparison of different group IV laser realizations and two group III/V lasers as references.

The status of group IV heterostructure lasers depends strongly on progress in the synthesis of metastable materials with defined strain values [51]. This can easily be seen by looking at the room temperature operation of these lasers. The basically hardest conditions are given by the very high electron densities of unstrained or slightly strained Ge, whereas the lower separation by the indirect-direct bandgap energy ΔE_c in highly strained Ge or GeSn requests much lower carrier injection levels. However, the experimental status reflects exactly the maturity of the material preparation and process stability, with the slightly strained Ge in the leading position.

Table 12.7 A selection of monolithic heterostructure lasers on Si

Active material	Direct semi-conductor	Wavelength (μm)	Electron stimu-lation	RT operation	Single laser line	CW	Ref.
GaInAlAs MQW	+	1.55	+	+	+	+	[42]
InAs, QD	+	1.3	–	+	+	+	[45]
n-Ge, unstrained	–	1.7	+	+	–	–	[49]
n-Ge, strained	—	1.6	—	+	—	—	[11]
n-Ge, strained	—	1.6	+	+	+	—	[12]
GeSn	+	2.3	—	—	—	—	[50]

Note: For comparison a hybrid laser on Si [42] is added.

12.8 Summary and Outlook

The lack of efficient Si light sources seemed to limit the applicability of Si-based photonics. Within a decade a variety of solutions have either already overcome or show the potential to overcome the small bottleneck in Si photonics. Solutions range from external light sources that couple their light via fiber in the chip, to hybrid lasers mounted on the chip, to a bundle of monolithic integration techniques (including III/V semiconductors and group IV semiconductors). The first two solutions offer good performances. Here the main focus is on production-friendly mounting techniques [52, 53]. The group IV lasers need the most intensive research efforts [41–62], with emphasis on strain, doping, Sn alloying, nanostructures, and epitaxy.

Excellent results were obtained for optically stimulated lasers at cryogenic temperatures [63–65]. High differential efficiencies

(<50%) were obtained with strained Ge microbridges [63]. The high strain of 5.4%–5.9% shifts the emission wavelength in the mid-IR region (3.2–3.7 μm). Lasing persists up to 100 K. Lasing persists nearly up to room temperature by combining GeSn alloys with high strain values [64]. Using tensile-strained microdisc GeSn resonators, low optical threshhold densities of about 1 kW/cm^2 could be obtained [65], allowing continuous wave operation up to 70 K. Obtaining room-temperature operation and electrical stimulation [66] of the laser will be dominant challenges for group IV laser research within the next few years.

References

1. N. Koshida (2007). Luminescence and related properties of nanocrystalline porous silicon, in *Landolt-Boernstein, New Series*, Vol. III, 34C3, Semiconductor Quantum Structures-Optical Properties (Springer-Verlag).

2. E. Kasper, H.-J. Herzog and H. Kibbel (1975). A one-dimensional SiGe superlattice grown by UHV epitaxy, *Appl. Phys.*, **8**, 199.

3. L. T. Canham (1990). Silicon quantum wire array fabrication by electrochemical and chemical dissolution of wafers, *Appl. Phys. Lett.*, **57**, 1046.

4. L. Pavesi, L. Dal Negro, C. Mazzoleni, G. Franzo and F. Priolo (2000). Optical gain in silicon nanocrystals, *Nature*, **408**, 440.

5. U. Gnutzmann and K. Clausecker (1974). Theory of direct optical transitions in an optical indirect semiconductor with a superlattice structure, *Appl. Phys.*, **3**, 9.

6. H. Brugger, G. Abstreiter, H. Jorke, H. J. Herzog and E. Kasper (1986). Dispersion of folded phonons in superlattices, *Phys. Rev. B*, **33**, 5928.

7. R. Zachai, K. Eberl, G. Abstreiter, E. Kasper and H. Kibbel (1990). Photoluminescence in short-period Si/Ge strained layer superlattices, *Phys. Rev. Lett.*, **64**, 1055.

8. H. Presting, U. Menzcigar and H. Kibbel (1993). Electro- and photoluminescence studies from ultrahin SimGen superlattices, *J. Vac. Sci. Technol. B*, **11**, 1110.

9. B. Corbett, C. Bower, A. Fecioru, M. Mooney, M. Gubbings and J. Justice (2013). Strategies for integration of lasers on silicon, *Semicond. Sci. Technol.*, **28**, 094001.

312 *Laser*

10. A. W. Fang, B. R. Koch, R. Jones, E. Lively, P. Liang, Y. K. Kuo and J. E. Bowers (2008). A distributed Bragg reflector silicon evanescent laser, *IEEE Photonics Technol. Lett.*, **20**, 1667.

11. J. Liu, X. Sun, R. Camacho-Aguilera, L. C. Kimerling and J. Michel (2010). Ge-on-Si laser operating at room temperature, *Opt. Lett.*, **35**, 679–681.

12. Jr. E. Camacho-Aguilera, Y. Cai, N. Patel, J. T. Bassette, M. Romangnoli, L. C. Kimerling and J. Michel (2012). An electrical pumped Ge laser, *Opt. Express*, **20**, 11316.

13. H. C. Casey and M. B. Panish (1978). *Heterostructure Lasers: Fundamental Principles* (Academic Press, New York).

14. C. Weisbuch (1987). Optical properties of quantum wells, in *Physics and Applications of Quantum Wells and Superlattices*, E. E. Mendez and K. von Klitzing, eds., NATO ASI Series B170 (Plenum Press), pp. 261–299.

15. M. Asada, Y. Miyamoto and Y. Suematsu (1986). Gain and the threshold of three-dimensional quantum box lasers, *IEEE J. Quantum Electron.*, **QE-22**, 1915.

16. W. T. Tsang (1984). Heterostructure semiconductor lasers prepared by MBE, *IEEE J. Quantum Electron.*, **QE-20**, 1119.

17. Y. Arakawa and A. Yariv (1985). Theory of gain, modulation response and spectral linewidth in AlGaAs quantum well lasers, *IEEE J. Quantum Electron.*, **QE-21**, 1666.

18. G. Mak and H. M. van Driel (1994). Femtosecond transmission spectroscopy at the direct band edge of germanium, *Phys. Rev. B*, **49**, 16817.

19. X. Q. Zhou, H. M. van Driel and G. Mak (1994). Femtosecond kinetics of photoexcited carriers in germanium, *Phys. Rev. B*, **50**, 5226.

20. S. A. Claussen, E. Tasyurek, J. E. Roth and D. A. B. Miller (2010). Measurement and modeling of ultrafast carrier dynamics and transport in germanium/silicon-germanium quantum wells, *Opt. Express*, **18**, 25596.

21. X. Wang, L. C. Kimerling, J. Michel and J. Liu (2013). Large inherent optical gain from the direct gap transition of Ge thin films, *Appl. Phys. Lett.*, **102**, 131116.

22. J. S. Blackmore (1982). Approximations for Fermi-Dirac integrals, *Solid State Electron.*, **25**, 1067.

23. S. M. Sze (1981). *Physics of Semiconductor Devices*, 2nd ed. (Wiley, New York).

24. E. Kasper, M. Oehme, J. Werner, T. Aguirov and M. Kittler (2012). Direct bandgap luminescence from Ge on Si pin diodes, *Front. Optoelectron.*, **5**, 256.

25. R. Neumann and G. Abstreiter (2000). Effective masses of electrons and holes in SiGe, in *EMIS Datareview Series*, Vol. 24 (INSPEC (IEE), London).

26. H. Kroemer (1978). The Einstein relation for degenerate carrier concentrations, *IEEE Trans.*, **ED-25**, 850.

27. E. Kasper and M. Oehme (2015). Germanium tin light emitters on silicon, *Jap. J. Appl. Phys.*, **54**, 04DG11.

28. R. Geiger, T. Zabel and H. Sigg (2015). Group IV direct bandgap photonics: methods, challenges and opportunities, *Front. Mater.*, **2**, 52.

29. J. Liu, R. Camacho-Aguilera, J. T. Bessette, X. Sun, S. Wang, Y. Cai, L. C. Kimerling and J. Michel (2012). Ge-on-Si optoelectronics, *Thin Solid Films*, **520**, 3354–3360.

30. G. Sun, R. A. Soref and H. H. Cheng (2010). Design of an electrically pumped SiGeSn/GeSn/SiGeSn double heterostructure mid-infrared laser diode, *Jpn. J. Appl. Phys.*, **108**, 033107.

31. T. Arguirov, M. Kittler, M. Oehme, N. V. Abrasimov, E. Kasper and J. Schulze (2011). Room temperature direct band gap emission from an unstrained Ge pin LED, *Solid State Phenom.*, **178**, 25.

32. D. S. Sukhdeo, S. Gupta, K. C. Saraswat, B. Dutt and D. Nam (2016). Impact of minority carrier lifetime on the performance of strained germanium light sources, *Opt. Commun.*, **364**, 233–237.

33. www.ioffe.ru/SVA/NSM/Semicond/Ge/Electronic.html

34. X. Sun (2009). Ge-on-Si light-emitting materials and devices for silicon photonics, PhD thesis, MIT.

35. J. Liu, X. Sun, D. Pan, X. Wang, L. C. Kimerling, T. L. Koch and J. Michel (2007). Tensile-strained, n-type Ge as a gain medium for monolithic laser integration on Si, *Opt. Express*, **15**, 11272.

36. P. Y. Yu and M. Cardona (1996). *Fundamentals of Semiconductors: Physics and Materials Properties* (Springer-Verlag, Berlin).

37. C. F. Klingshirn (2012). *Semiconductor Optics*, 4th ed. (Springer-Verlag, Berlin).

38. C. Xu, J. D. Gallagher, C. L. Senaratne, J. Menendez and J. Kouvetakis (2016). Optical properties of Ge-rich $Ge_{1-x}Si_x$ alloys: compositional dependence of the lowest direct and indirect gaps, *Phys. Rev. B*, **93**, 125206.

39. P. H. Lim, S. Park, Y. Ishikawa and K. Wada (2009). Enhanced direct bandgap emission in germanium by micromechanical strain engineering, *Opt. Express*, **17**, 16358.

40. S. Fathpour (2012). Emerging heterogeneous integrated platforms on silicon, *Nanophotonics*, **4**(1).

41. A. Spott, M. Davenport, J. Peters, J. Bovington, M. J. R. Heck, E. J. Stanton, I. Vurgaftman, J. Meyer and J. E. Bowers (2015). Heterogeneously integrated 2.0 μm CW hybrid silicon lasers at room temperature, *Opt. Lett.*, **40**, 1480.

42. C. Zhang, S. Srinivasan, Y. Tang, M. R. J Heck, M. L. Davenport and J. E. Bowers (2014). Low threshold and high speed short cavity distributed feedback hybrid silicon lasers, *Opt. Express*, **22**, 10202–10209.

43. H. T. Hattori, C. Seassal, E. Touraille, P. Rojo-Romeo, X. Letartre, G. Hollinger, P. Viktorovitch, L. Di Cioccio, M. Zussy, L. El Melhaoui and J. M. Fedeli (2006). Heterogeneous integration of microdisk lasers on silicon strip waveguides for optical interconnects, *IEEE Photonics Technol. Lett.*, **18**, 223.

44. S. Keyvaninia, C. Roelkens, D. Van Thourhout, C. Jany, M. Lamponi, A. L. Liepvre, F. Lelarge, D. Make, G. H. Duan, D. Bordel and J. M. Fedeli (2013). Demonstration of a heterogeneously integrated III-V/SOI single wavelength tuneable laser, *Opt. Express*, **38**, 5434.

45. Y. Wan, Q. Li, A. Y Liu, A. C. Gossard, J. E. Bowers, E. L. Hu and K. M. Lau (2016). Temperature characteristics of epitaxially grown InAs quantum dot micro-disk lasers on silicon for on-chip light sources, *Appl. Phys. Lett.*, **109**, 011104.

46. A. Y. Liu, C. Zhang, J. Norman, A. Snyder, D. Lubyshev, J. M. Fastenau, A. W. K. Liu, A. C. Gossard and J. E. Bowers (2014). High performance continuous wave 1.3 μm quantum dot laser on silicon, *Appl. Phys. Lett.*, **104**, 041104.

47. Z. Zhau, B. Yin and J. Michel (2015). On-chip light sources for silicon photonics, *Light Sci. Appl.*, **4**, e358.

48. L. Carroll, P. Friedli, S. Neuenschwander, H. Sigg, S.Cecchi, F. Isa, D. Chrastina, G. Isella, Y. Fedoryshyn and J. Faist (2012). Direct-gap gain and optical absorption in germanium correlated to the density of photo exited carrier, doping and strain, *Phys. Rev. Lett.*, **109**, 057402.

49. R. Koerner, M. Oehme, M. Gollhofer, M. Schmid, K. Kostecki, S. Bechler, W. Widmann, E. Kasper and J. Schulze (2015). Electrically pumped lasing from Ge Fabry-Perot resonators on Si, *Opt. Express*, **23**, 14815.

50. S. Wirths, R. Geiger, N. von den Driesch, G. Mussler, T. Stoica, S. Mantl, Z. Ikonic, M. Luysberg, S. Chiussi, J. M. Hartmann, H. Sigg, J. Faist, D. Buca and D. Grützmacher (2015). Lasing in direct-bandgap GeSn alloy grown on Si, *Nat. Photonics*, **9**, 88–92.

51. E. Kasper (2016). Group IV heteroepitaxy on Si for photonics, *J. Mater. Res.*, **31**, 3639–3649.

52. E. Kasper and V. Stefani (2008). Device with optical coupling window and fabrication method for this device, German patent, No. 102008051625 (in German).

53. J. Zhang, B. Haq, J. O'Callaghan, A. Gocalinska, E. Peluchi, A. J. Trinade, B. Corbette, G. Morthier and G.Roelkens (2018). Transfer-printing-based integration of a III-V-on-silicon distributed feedback laser, *Opt. Express*, **26**, 8821–8830.

54. B. Dutt, D. S. Sukhdeo, D. Nam, B. M. Vulovic, Z. Yuan and K. C. Saraswat (2012). Roadmap to an efficient germanium-on-silicon laser: strain vs. n-type doping, *IEEE Photonics J.*, **4**, 2002.

55. Z. Zhou, B. Yin and J. Michel (2015). On-chip light sources for silicon photonics, *Light Sci. Appl.*, **4**, e358.

56. Y. Cai, Z. Han, X. Wang, R. E. Camacho-Aguilera, L. C. Kimerling, J. Michel and J. Liu (2013). Analysis of threshold current behavior for bulk and quantum well germanium laser structures, *IEEE J. Sel. Top. Quantum Electron.*, **19**, 1901009.

57. D. Stange, S. Wirths, N. von den Driesch, G. Mussler, T. Stoica, Z. Ikonic, J. M. Hartmann, S. Mantl, D. Grützmacher and D. Buca (2015). Optical transitions in direct-bandgap $Ge_{1-x}Sn_x$ alloys, *ACS Photonics Lett.*, **2**, 1539–1545.

58. V. Reboud, et al. (2018). Optically pumped GeSn micro-disks with 16% Sn lasing at 3.1 µm up to 180 K, *Appl. Phys. Lett.*, **111**, 092101.

59. R. Loo, Y. Shimura, S. Ike, A. Vohra, T. Stoica, D. Stange, D. Buca, D. Kohen, J. Margetis and J. Tolle (2018). Epitaxial GeSn: impact of process conditions on material quality, *Semicond. Sci. Technol.*, **33**, 114010 (9pp).

60. N. von den Driesch, et al. (2018). Advanced GeSn/SiGeSn group IV heterostructure lasers, *Adv. Sci.*, **5**, 1700955.

61. P. O. Vaccaro, M. I. Alonso, M. Garriga, J. Gutiérrez, D. Peró, M. R. Wagner, J. S. Reparaz, C. M. Sotomayor Torres, X. Vidal, E. A. Carter, P. A. Lay, M. Yoshimoto and A. R. Goñi (2018). Localized thinning for strain concentration in suspended germanium membranes and optical method for precise thickness measurement, *AIP Adv.*, **8**, 115131.

62. G. Niu, G. Capellini, M. A. Schubert, T. Niermann, P. Zaumseil, J. Katzer, H.-M. Krause, O. Skibitzki, M. Lehmann, Y.-H. Xie, H. von Känel and T. Schroeder (2016). Dislocation-free Ge nano-crystals via pattern independent selective Ge heteroepitaxy on Si nano-tip wafers, *Sci. Rep.*, **6**, 22709.

63. F. T. Armand Pilon et al. (2019). Lasing in strained germanium microbridges, Nat. Commun., 10, 2724.

64. J. Chrétien et al. (2019). GeSn lasers covering a wide wavelength range thanks to uniaxial tensile strain, ACS Photon., 6, 2462–2469.

65. A. Elbaz et al. (2020). Ultra-low-threshold continuous-wave and pulsed lasing in tensile-strained GeSn alloys, Nat. Photon., arXiv:2001.04927.

66 Y. Zhou et al. (2020). Electrically injected GeSn lasers on Si operating up to 100 K, arXiv 2004.09402.

Chapter 13

Future Challenges

Silicon photonics is now on the verge of entering the mass market from a niche one. Optimistic assessments [1] of the market forecast a large annual growth rate (more than 40%) up to a volume of around 0.5 billion US dollars at the chip level and almost 4 billion US dollars at the transceiver level in 2025. More than 80% of this volume is assigned to optical intra-data-center connects. Indeed, data centers could be the market drivers for silicon photonics, with their need for low cost per data lane, low power consumption per data lane, high reliability, and high fabrication yield. Other market segments with high commercialization potential are telecommunication (metropolitan areas) and medical equipment. Silicon photonics delivers key components to aeronautics/aerospace, sensors, autonomous traffic, and high-performance computing. The tipping point for silicon-based photonics was obtained by adding Ge-on-Si photodetectors to CMOS technology on SOI substrates. This allowed complete monolithic integration of the receiver part, but silicon photonics is at a lower level of maturity than the electronics industry, and there are still challenges to overcome. For these challenges, technical breakthroughs will be necessary in laser source performance, in small size and small power modulators, and in monolithic integration techniques.

Silicon-Based Photonics
Erich Kasper and Jinzhong Yu
Copyright © 2020 Jenny Stanford Publishing Pte. Ltd.
ISBN 978-981-4303-24-8 (Hardcover), 978-981-4303-25-5 (eBook)
www.jennystanford.com

13.1 Group IV Laser Performance

The largest principal challenge concerns group IV lasers and efficient light sources (light-emitting diode [LED]), which are monolithically integrated on silicon. Basic research on this topic within the last decade has given two very remarkable results.

- Optical gain and electrically induced laser emission from band-to-band transitions are possible even with indirect semiconductors if the energetic distance between the indirect and direct bandgaps is small. Electrically stimulated laser emission from slightly strained and unstrained Ge on Si (120–136 meV energy difference between direct and indirect transitions) confirmed room temperature operation [2, 3], although with high threshold currents.

- Direct bandgaps with group IV semiconductors are possible with highly tensile strained Ge (biaxial tensile strain of about 2% and uniaxial strain of more than 4%) or GeSn alloys (Sn amount of more than 7%). Both structures are only metastable at the thicknesses needed for optical devices. Optically stimulated emission [4] is demonstrated at low temperatures. Both mentioned directions of material modification toward the direct optical character of group IV semiconductors shift the emission wavelengths more in the infrared (>1.6 µm) region compared to Ge.

The rather slow progress of group IV laser device realization is because of quality problems in layer preparation and device processing of metastable materials [5]. Narrow process windows are necessary to avoid strain relaxation of highly strained structures from misfit dislocations and to avoid phase separation from homogenous GeSn alloys containing more than 1% Sn. The main measure of the material quality of a laser structure is given by the nonradiative recombination rate. Nonradiative recombination via midgap defect levels (contaminants, dislocations, vacancies, and interstitials) competes with the radiative recombination of the laser. The group IV laser question motivated a broad, basic investigation of properties and stability of metastable materials, which is interesting in itself and for its high application potential. Sophisticated defect characterization methods [6], like deep-level transient spectroscopy, polarized Raman spectroscopy [7], and atom probe tomography [8]; surface studies of the phase separation [9] at the limit of metastability; and investigations of the strain relaxation mechanism and its effects on Sn incorporation [10], gave valuable insight into the complex interplay of different growth parameters. Insight into the growth process resulted in suggestions of virtual higher-quality substrates [11]. A high density of misfit dislocations (typical dislocation spacing of 10 nm) at the interface to the Si substrate is necessary for elastic strain relaxation. However, growth kinetics allow the densities of the undesired threading dislocations (TDs) to vary by several orders of magnitude. In Ref. [11], they modeled the reduction of TD density in heteroepitaxial coalesced layers in terms of the bending of TDs induced by image forces at nonplanar selective epitaxial growth surfaces before the coalescence. The reduction of TD density was quantitatively verified for Ge layers on (001) Si with line-and-space SiO_2 masks. GeSn layers (Sn contents from 5% to 15%) on virtual Ge substrates have a lattice mismatch of about 0.7% to 2.2%. The strain relaxation in such metastable GeSn is investigated under low-temperature growth conditions for different amounts of Sn (up to 5% [12], up to 13% [13], and up to 22% [14]). Alternative growth strategies use lateral growth of GeSn [15] directly on Si oriented in the (111) direction. GeSn layers show good thermal stabilities at annealing temperatures of up to 500°C [15]. A Fabry–Perot-type GeSn waveguide with 12.6% Sn demonstrated lasing with an optical pumping threshold density of 325 kW cm^{-2} at 20 K [4], and a couple of more successful demonstrations have since followed. However, the lasing threshold of those GeSn lasers was still too high to be practical, possibly owing to faster-than-desired nonradiative recombination processes. On the other hand, the application of large mechanical tensile strain [16] can also address the challenge of the indirect bandgap in Ge by fundamentally altering the band structure; the small energy difference between the Γ and L conduction valleys can be reduced even further with tensile strain, resulting in an increased material gain. With regard to practical realization, several innovative platforms for inducing large mechanical strain have been experimentally demonstrated [17]. The status of laser development is explained on the well-documented example of a uniaxially strained Ge microbridge [16] that is in stiction with the underlying oxide layer. The close contact improves the thermal resistance, which is the

320 | *Future Challenges*

thermal bottleneck in suspended microbridges. Multimode lasing from pulsed optical pumping was observed up to a temperature of 83 K. A pair of distributed Bragg reflectors provided the light amplification in the gain medium (strained Ge bridge). The onset of lasing was determined by the visibility of narrow multimode peaks and by the superlinear increase of intensity. This usually chosen procedure in low-power emitters simplifies the role of bleaching before the onset of net gain. The losses caused mainly by free carrier absorption are partly overcome by the gain [18], leading to a net absorption bleaching below the laser onset. In strained Ge, the intervalence band absorption from free holes is assumed to be larger as the free electron contribution because of hole transitions between the valence sub-band's light/heavy hole and spin off hole.

The low-dimensional structure may be designed by means of lithography, as in the given example, which is a technologically advanced method. However, bottom-up nanowires, where the constituent atoms assemble to form a larger structure [19], will be under focus in future work. Low-dimensional core/shell nanowire structures demonstrated good optical emission and absorption properties [20]. Design works focus on GeSnSi/GeSn quantum wells [21] for proper carrier confinement in diode structures or transistor laser structures [22].

13.2 Monolithic Integration Issues

A less spectacular but essential challenge is given by the need for integration of different photonic devices and for the integration of photonic integrated circuits with conventional microelectronic circuits [23]. Simply speaking, the use of a common substrate material is only one important prerequisite of monolithic integration. Also, the different device structures and material properties have to be similar enough [24] that only a few additional fabrication steps have to be added. Let us give an example from microelectronics to outline the importance of these aspects. A combination of high-transconductance bipolar circuits with complementary metal-oxide-semiconductor (CMOS) logic—called BICMOS—offers advantages for highly desirable system integration of digital processing and input/output units. Both approved technologies utilize silicon substrates and silicon manufacturing. The market share of BICMOS is in the lower-percentage regime despite minimal technical problems because of necessary trade-offs in device optimization. Si-based photonics use diode-type devices for the detection, modulation, and emission of light [25]. Detection is optimized for low-power zero-bias operation with a low-doped absorption region; the corresponding absorption wavelength is in the energy level above the bandgap. Absorption modulators are working with the same diode structure [26], which is switched between two reverse-voltage values for signal modulation. However, the modulation wavelength corresponds to an energy slightly below the bandgap energy. LED and laser are operated under forward bias, and the emission energy is above the bandgap energy. In principle, the simple diode structure is again usable. However, optimization for good carrier confinement tends toward low-dimensional structures within the emission region [27]. Indeed, monolithic photonic integration addresses research topics [28] like system architecture, device physics, fabrication techniques, and material synthesis. The chosen wavelength regime influences essentially the system architecture. For the telecommunication wavelength regime from 1.3 μm to 1.7 μm, the principal scheme will be dominated by external light sources that couple in light via fiber connections into the on-chip photonic waveguide network. Actually, that is the existing procedure in laboratory work, which allows the use of the sophisticated discrete group III–V lasers and the highly developed telecommunication measurement equipment. Substantial improvements are needed for self-aligned mounting and housing techniques, which are the bottleneck for large-scale use. Ideally, different optical input/output configurations are available, similar to the situation with electrical power and signals. On a chip, the silicon waveguide network is supported by Ge-based detectors and modulators. Ge-based detectors are already accepted in manufacturing [23]. Ge-based absorption modulators are assumed to outperform the interference modulators with respect to low power consumption [24]. Integrated light sources will be in our opinion the exception in this wavelength regime, although excellent research results from monolithic and hybrid integrated group III–V compounds were reported and although room temperature operation of Si-based Ge lasers with electrical stimulation was demonstrated. Complex integration schemes for group III–V devices [29, 30] and high threshold currents for Ge-on-Si lasers are the reasons for this

Future Challenges

assessment. We predict a different situation for the wavelength region of 2–4 μm, with a high potential for the integration of light sources based on GeSn alloys or strained Ge [31]. At first glance, this assessment may astonish because neither room temperature operation nor electrical laser stimulation was demonstrated but both were expected due to the conversion of the indirect semiconductor Ge into a direct one by applying tensile elastic strain or alloying with Sn. In both cases the material is only metastable but strong progress is seen in the structural stability and preparation of such material classes [5]. In the future, the electrical properties [32] have to be given higher priority to understand the device behavior. A p-type background was detected by investigations of Zaima and group [33], which could be a hint for vacancy concentrations (more than $10^{16}/cm^2$) much higher than usual in Si. Differential reflectivity spectroscopy [34] emerged as a versatile method to judge the recombination properties of carriers in as-grown layers. Common structural, optical, and electrical investigations push the search for higher performance of laser devices from metastable materials.

Some research fields, now of high scientific interest, will rapidly gain more practical importance with device scaling in integrated Si-based photonic circuits. These fields include plasmonics [35, 36], photonic crystals [37], nonlinear effects [38, 39], and millimeter-wave generation [40, 41] from light mixing.

References

1. E. Mounier and J.-L. Malinge (2018). Yole annual report, http://www.yole.fr/Silicon Photonics_Market_Applications.aspx

2. E. Camacho-Aguilera, Y. Cai, N. Patel, J. T. Bassette, M. Romangnoli, L. C. Kimmerling and J. Michel (2012). An electrical pumped Ge laser, *Opt. Express*, **20**, 11316.

3. R. Koerner, M. Oehme, M. Gollhofer, M. Schmid, K. Kostecki, S. Bechler, W. Widmann, E. Kasper and J. Schulze (2015). Electrically pumped lasing from Ge Fabry-Perot resonators on Si, *Opt. Express*, **23**, 14815.

4. S. Wirths, R. Geiger, N. von den Driesch, G. Mussler, T. Stoica, S.Mantl, Z. Ikonic, M. Luysberg, S. Chiussi, J. M. Hartmann, H. Sigg, J. Faist, D. Buca and D. Grützmacher (2015). Lasing in direct-bandgap GeSn alloy grown on Si, *Nat. Photonics*, **9**, 88–92.

5. E. Kasper (2016). Group IV heteroepitaxy on silicon for photonics, *J. Mater. Res.*, **31**, 3639–3648.

6. S. Gupta, E. Simoen, R. Loo, Y. Shimura, C. Porret, F. Gencarelli, K. Paredis, H. Bender, J. Lauwaert, H. Vrielinck and M. Heyns (2018). Electrical properties of extended defects in strain relaxed GeSn, *Appl. Phys. Lett.*, **113**, 022102.

7. T. S. Perova, E. Kasper, M. Oehme, S. Cherevkov and J. Schulze (2017). Features of polarized Raman spectra for homogeneous and non-homogeneous compressively strained GeSn alloys, *J. Raman Spectrosc.*, **48**, 993–1001.

8. S. Assali, J. Nicolas, S. Mukherjee, A. Dijkstra and O. Moutanabbir (2018). Atomically uniform Sn-rich GeSn semiconductors with 3.0–3.5 μm room-temperature optical emission, *Appl. Phys. Lett.*, **112**, 251903.

9. L. Kormos, M. Kratzer, K. Kostecki, M. Oehme, T. Sikola, E. Kasper, J. Schulze and C. Teichert (2017). Surface analysis of epitaxially grown GeSn alloys with Sn contents between 15% and 18%, *Surf. Interface Anal.*, **49**, 297–302.

10. W. Dou, M. Benamara, A. Mosleh, J. Margetis, P. Grant, Y. Zhou, S. Al-Kabi, W. Du, J. Tolle, B. Li, M. Mortazavi and S.-Q. Yu (2018). Investigation of GeSn strain relaxation and spontaneous composition gradient for low-defect and high-Sn alloy growth, *Sci. Rep.*, **8**, 5640.

11. M. Yako, Y. Ishikawa, E. Abe and K. Wada (2018). Reduction of threading dislocations by image force in Ge selective epilayers on Si, *Proc. SPIE 10823, Nanophotonics and Micro/Nano Optics IV*, 108230F, doi:10.1117/12.2501081.

12. K. R. Khiangte, J. S. Rathore, V. Sharma, S. Bhunia, S. Das, R. S. Fandan, R. S. Pokharia, A. Laha and S. Mahapatra (2017). Dislocation density and strain-relaxation in $Ge_{1-x}Sn_x$ layers grown on Ge/Si (001) by low-temperature molecular beam epitaxy, *J. Cryst. Growth*, **470**, 135–142.

13. V. P. Martovitsky, Yu. A. Aleshchenko, V. S. Krivobok, A. V. Muratov, A. V. Klekovkin and A. B. Mehiya (2018). Molecular beam epitaxy of Si–Ge–Sn heterostructures for monolithically integrated optoelectronic devices based on silicon, *Bull. Russ. Acad. Sci.*, **82**, 418–423.

14. W. Dou, M. Benamara, A. Mosleh, J. Margetis, P. Grant, Y. Zhou, S. Al-Kabi, W. Du, J. Tolle, B. Li, M. Mortazavi and S.-Q. Yu (2018). Investigation of GeSn strain relaxation and spontaneous composition gradient for low-defect and high-Sn alloy growth, *Sci. Rep.*, **8**, 5640.

15. D. Zhang, Z. Liu, D. Zhang, X. Zhang, J. Zhang, J. Zheng, Y. Zuo, C. Xue, C. Li, S. Oda, B. Cheng and Q. Wang (2018). Sn-guided defect-free GeSn

lateral growth on Si by molecular beam epitaxy, *J. Phys. Chem. C*, **119**, 17842–17847.

16. S. Bao, D. Kim, C. Onwukaeme, S. Gupta, K. Saraswat, K. H. Lee, Y. Kim, D. Min, Y. Jung, H. Qiu, H. Wang, E. A. Fitzgerald, C. S. Tan and D. Nam (2017). Low-threshold optically pumped lasing in highly strained germanium nanowires, *Nat. Commun.*, **8**, 1845, doi: 10.1038/s41467-017-02026.

17. R. A. Minamisawa, M. J. Süess, R. Spolenak, J. Faist, C. David, J. Gobrecht, K. K. Bourdelle and H. Sigg (2012). Top-down fabricated silicon nanowires under tensile elastic strain up to 4.5%, *Nat. Commun.*, **3**, 1096, doi: 10.1038/ncomms2102.

18. R. Koerner, M. Oehme, M. Gollhofer, K. Kostecki, M. Schmid, S. Bechler, D. Widmann, E. Kasper and J. Schulze (2014). *Proc. 7th International Silicon-Germanium Technology and Device Meeting (ISTDM)*, Singapore, 121–122.

19. M. Amato, M. Palummo, R. Rurali and S. Ossicini (2014). Silicon–germanium nanowires: chemistry and physics in play, from basic principles to advanced applications, *Chem. Rev.*, **114**, 1371–1412.

20. S. Assali, A. Dijkstra, A. Li, S. Koelling, M. A. Verheijen, L. Gagliano, N. von den Driesch, D. Buca, P. M. Koenraad, J. E. M. Haverkort and E. P. A. M. Bakkers (2017). Growth and optical properties of direct band gap Ge/Ge0.87Sn0.13 core/shell nanowire arrays, *Nano Lett.*, **17**, 1538–1544.

21. Y. Shimura, S. A. Srinivasan and R. Loo (2016). Design requirements for group-IV laser based on fully strained $Ge_{1-x}Sn_x$ embedded in partially relaxed $Si_{1-y-z}Ge_ySn_z$ buffer layers, *ECS J. Solid State Sci. Technol.*, **5**, Q140–Q143.

22. B. Mukhopadhya, G. Sen, S. De, R. Basu, V. Chakraborty and P. K. Basu (2018). Calculated characteristics of a transistor laser using alloys of Gr-IV elements, *Phys. Status Solidi B*, 1800117.

23. G.-S. Jeong, W. Bae and D.-K. Jeong (2017). Review of CMOS integrated circuit technologies for high-speed photo-detection, *Sensors*, **17**, 1962.

24. X. Wang and J. Liu (2018). Emerging technologies in Si active photonics, *J. Semicond.*, **39**, 061001, doi: 10.1088/1674-4926/39/6/061001.

25. M. Oehme, E. Kasper and J. Schulze (2013). GeSn heterojunction diode: detector and emitter in one device, *ECS J. Solid State Sci. Technol.*, **2**, 76–78.

26. M. Oehme, K. Kostecki, M. Schmid, M. Kaschel, M. Gollhofer, K. Ye, D. Widmann, R. Koerner, S. Bechler, E. Kasper and J. Schulze (2014). Franz-Keldysh effect in GeSn pin photodetectors, *Appl. Phys. Lett.*, **104**, 161115.

27. M. Oehme, D. Widmann, K. Kostecki, P. Zaumseil, B. Schwartz, M. Gollhofer, R. Koerner, S. Bechler, M. Kittler, E. Kasper and J. Schulze (2014). GeSn/Ge multi quantum well photodetectors on Si substrates, *Opt. Lett.*, **39**, 4711–4714.

28. X. Chen, M. M. Milosevic, S. Stankovic, S. Reynolds, T. D. Bucio, K. Li, D. J. Thompson, F. Gardes and G. T. Reed (2018). The emergence of silicon photonics as a flexible technology platform, *Proc. IEEE*, doi: 1.11090/ JPROC.2018.2854372.

29. K. Sun, D. Jung, C. Shang, A. Liu, J. Morgan, J. Zang, Q. Li, J. Klamkin, J. E. Bowers and A. Belin (2018). Low dark current III-V on silicon photodiodes by heteroepitaxy, *Opt. Express*, **26**, 13605.

30. J. Zhang, B. Haq, J. O'Callaghan, A. Gocalinska, E. Peluchi, A. J. Trinade, B. Corbette, G. Morthier and G. Roelkens (2018). Transfer-printing-based integration of a III-V-on-silicon distributed feedback laser, *Opt. Express*, doi: 10.1364/OE.26.008821.

31. Z. Qi, H. Sun, M. Luo, Y. Jung and D. Nam (2018). Strained germanium nanowire optoelectronic devices for photonic integrated circuits, *J. Phys.: Condens. Matter*, **30**, 334004.

32. S. Gupta, E. Simoen, R. Loo, Y. Shimura, C. Porret, F. Gencarelli, K. Paredis, H. Bender, J. Lauwaert, H. Vrielinck and M. Heyns (2018). Electrical properties of extended defects in strain relaxed GeSn, *Appl. Phys. Lett.*, **113**, 022102, doi: 10.1063/1.5034573.

33. O. Nakatsuka, N. Tsutsui, Y. Shimura, S. Takeuchi, A. Sakai and S. Zaima (2010). Mobility behavior of $Ge_{1-x}Sn_x$ layers grown on silicon-on-insulator substrates, *Jpn. J. Appl. Phys.*, **49**, 04DA10.

34. P. Onufrijs, personal information.

35. D. V. Guzatov, S. V. Gaponenko and H. V. Demir (2018). Plasmonic enhancement of electroluminescence, *AIP Adv.*, **8**, 015324.

36. M. Gu, P. Bai, H. S. Chu and E.-P. Li (2012). A design of subwavelength CMOS compatible plasmonic photodetector for nano-photonic integrated circuits, *IEEE Photonics Technol. Lett.*, **24**, 515–517.

37. P. Cheben, R. Halir, J. H. Schmid, H. A. Atwater and D. R. Smith (2018). Subwavelength integrated photonics, *Nature*, **560**, 565–572.

38. L. Zhang, A. M. Agarwal, L. C. Kimerling and J. Michel (2014). Nonlinear group IV photonics based on silicon and germanium: from near-infrared to mid-infrared, *Nanophotonics*, **3**, 247–268.

39. R. Osgood Jr., J. B. Driscoll, W. Astar, X. Liu, J. I. Dadap, W. M. J. Green, Y. A. Vlasov and G. M. Carter (2010). Nonlinear silicon photonics, *SPIE Newsroom*, doi: 10.1117/2.1201004.002934.

40. B. Yoo, R. P. Scott, D. J. Geisler, N. K. Fontaine and F. M. Soares (2012). Terahertz information and signal processing by RF-photonics, *IEEE Trans. Terahertz Sci. Technol.*, **2**, 167–176.

41. A. W. M. Mohammad (2019). Integrated photonics for millimetre wave transmitters and receivers, PhD thesis, University College London, https://www.researchgate.net/publication/334082977.

Index

absorption, 16–17, 179, 182–85, 187–88, 190, 192, 197, 199–200, 202, 226, 233, 236–38, 258–60, 274–76, 302–4
  direct, 304
  direct band-to-band, 304
  free hole, 306
  fundamental spontaneous, 302
  indirect, 185
  indirect band, 304
  interband, 15–16
  intervalence band, 320
  intrinsic, 39
  near-band-edge, 208
  steep, 186
  stimulated, 292
  two-photon, 85
absorption coefficients, 14, 16–17, 72, 184, 193, 208, 228, 230, 232–34, 236–37, 258, 303–4
  free carrier, 306
active pixel sensor (APS), 182
add-drop filters, 51, 64, 81–84, 245
all-pass filter (APF), 64, 81–82, 88–89, 147–51
Anderson model, 196
APDs, *see* avalanche photodetectors
APF, *see* all-pass filter
APS, *see* active pixel sensor
Auger effect, 212–13, 300
Auger process, 300–301
Auger recombination, 168, 195, 204, 293, 300
Avalanche photodetectors (APDs), 199, 211

back end of the line (BEOL), 225
band edges, 17, 62–63, 152–53, 169–70, 208–9, 216, 286, 288, 292, 304, 306
  conduction, 170, 282
  direct, 287, 305
  indirect, 287
  indirect conduction, 296
  steep, 83
bandgaps, 10, 17–19, 126–27, 152–53, 158–59, 168–70, 172–73, 175, 185–88, 226, 230, 251–53, 258, 260–61, 263–66
bandgap semiconductors, 200–201, 261
  indirect, 234
beam propagation method (BPM), 31, 33, 35–36
BEOL, *see* back end of the line
BER, *see* bit-error rate
biaxial stress, 175, 264–65, 294
birefringence, 45–46
bit-error rate (BER), 211
BL, *see* buried layer
Bloch boundary conditions, 151
Bloch form, 126
Bloch theory, 126
Bloch wave number, 152
Bloch waves, 137, 153
Bloch wave vector, 152
Boltzmann approximation, 174, 194, 285–87, 290–91
Boltzmann constant, 194
Boltzmann distribution, 274
BPM, *see* beam propagation method
Bragg cell constructs, original, 127

328 | *Index*

Bragg condition, 110–11
Bragg frequency, 152
Bragg gap, 152
Bragg period, 152
Bragg reflection, 114, 152
Bragg-type process, 152
Brillouin zone (BZ), 10, 127,
    157–58, 167–68, 171, 173,
    185–86, 272, 289, 295
buried layer (BL), 174–75, 197
bus waveguides, 50–52, 54, 61–62,
    81–82, 85
butt coupling, 117, 210, 248
Butterworth filter, 62
BZ, *see* Brillouin zone

carrier density, 195, 283, 285,
    291–92, 294–96, 298–99, 302,
    305–6
carrier injections, 173, 217, 232,
    237, 245, 271, 275, 278, 284,
    302
carriers, 167, 169, 172, 186,
    194–95, 197, 199, 212–13,
    238, 240, 243, 245, 273–76,
    281, 283–84
cascaded microrings, 52–53,
    56–57, 61–64, 88, 148
    large-scale, 154
cavity modes, 138, 276–77
Chebyshev filters, 62
chemical vapor deposition, 78, 262
    low-pressure, 78
    plasma-enhanced, 78, 116
cladding, 2, 4, 24, 26–28, 39,
    41–42, 45, 71, 73, 95–96,
    104–5, 107–8, 116, 130, 255
CMOS, *see* complementary metal-
    oxide-semiconductor
complementary metal-oxide-
    semiconductor (CMOS), 49,
    73, 95, 105, 109, 113–15, 146,
    181–82, 201, 247, 317, 320

conduction bands, 10–11, 15, 18,
    167–68, 170–71, 173, 185–86,
    195–96, 200, 202, 263–66,
    278–80, 285–87, 290, 292
    direct, 18, 174, 285, 292, 294,
    298
    indirect, 284–85, 298
    indirect/direct, 265
continuous wave (CW), 302, 311
coplanar electrical waveguide
    (CPW), 182–83
coupled-resonator optical
    waveguide (CROW), 88–89,
    148–49, 153–55
coupling, 51–52, 62, 68, 70–71, 79,
    81–82, 96–101, 103, 105–8,
    111–14, 117, 129–30, 138–40,
    149–54, 245
    critical, 58, 81, 149
    direct, 148
    evanescent, 210, 248, 272, 308
    low-loss, 35
    symmetric, 65
    symmetrical, 67
    vertical, 112, 114
CPW, *see* coplanar electrical
    waveguide
CROW, *see* coupled-resonator
    optical waveguide
CW, *see* continuous wave

dark currents, 196, 198, 204–7,
    211, 215–18
deep ultraviolet (DUV), 74–75
demultiplexer/multiplexer, 83
    optical, 81
depletion layer, 193, 195–98,
    212–13, 256–57
    voltage-dependent, 256
detectors, 138, 179–80, 184, 186,
    188, 190, 192, 194–96, 198,
    200–202, 204, 206, 208–10,
    212–20, 222

bulk silicon, 201
low-bandgap, 179
millimeter-wave-frequency-modulated, 195
photocurrent-based, 195
photodiode, 190
thermal, 179–80
vertical, 214
zero-bias, 215
DFB, *see* distributed feedback
DH, *see* double heterostructure
diamond lattice, 9–10, 15, 167, 171–72
centrosymmetric, 85
cubic, 185
dielectric function, 12–15
diffraction, 22, 108
first-order, 110–11
second-order, 110
upward, 113
diffusion, 101, 184, 194–95, 212–13, 237, 256, 299
slow, 213
slow carrier, 212
slow minority carrier, 212
digital pixel sensor (DPS), 182
direct bandgap semiconductors, 86, 251, 303
direct bandgap transitions, 218, 230, 258–59, 261, 303
direct semiconductors, 11–12, 172, 185, 208, 261, 271, 278, 285, 295, 298, 310
distributed feedback (DFB), 308
doping, 175, 189, 192–94, 199, 211–13, 217, 256, 281, 296, 302–3, 308, 310
high contact, 204
n-type, 175
optimum, 295, 297
double heterostructure (DH), 281–83, 307
double microrings, 88

coupled, 63
parallel-coupled, 88
DPS, *see* digital pixel sensor
Drude model, 306
DUV, *see* deep ultraviolet

EAMs, *see* electroabsorption modulators
EBL, *see* electron beam lithography
effective index method (EIM), 30–31, 36–37, 71–72
EIM, *see* effective index method
Einstein coefficients, 274
Einstein condensates, 146
EIT, *see* electromagnetically induced transparency
electroabsorption modulators (EAMs), 226, 230, 236, 247
electromagnetically induced transparency (EIT), 88
electron beam lithography (EBL), 74–76, 105, 140–41
electro-optic effect, 80, 227, 229–30, 240, 247
linear, 235
electro-optic modulators, 234, 236, 238–39, 247
high speed, 245
silicon-based, 236
emission, 1, 4, 136, 168–69, 174–75, 188–89, 259, 273–75, 277, 301, 303, 309, 311, 318, 320–21
etching, 2–3, 40, 68–69, 72–73, 76–78, 106–7, 109, 113, 116, 131–32, 271, 308
anisotropic, 76
chemical, 77
dry, 2, 74, 76–77, 101
reactive ion, 107
selective, 3
stepped, 101
wet, 76

exclusive NOR (XNOR), 87–88

Fabry–Perot cavity mode
  spectrum, 276
Fabry–Perot resonator, 51–52, 86,
  88, 236, 243, 277
Fabry–Perot ridge waveguide, 275
far infrared (FIR), 180, 200–201
FDTD, *see* finite difference time
  domain
FEOL, *see* front end of the line
Fermi–Dirac statistics, 170, 194,
  278–79, 284–85, 287
Fermi–Direc integral, 285–86
Fermi energy, 278, 287, 292
Fermi level, 196, 211, 306
FIB, *see* focused ion beam
figure of merit (FOM), 85–86
filters, 50, 62, 81, 83, 135, 139,
  142, 201
  optical, 61, 80–81, 89
  tunable, 81, 84–85
finite difference time domain
  (FDTD), 31, 127–28, 137,
  139–40
FIR, *see* far infrared
FKE, *see* Franz–Keldish effect
focused ion beam (FIB), 115
FOM, *see* figure of merit
four-wave mixing (FWM), 85
Franz–Keldish effect (FKE), 208–9,
  226–27, 230–31, 234–36
free carrier absorption, 39, 283,
  293, 302, 304, 306, 309, 320
free carrier dispersion effect, 84,
  87
free-spectrum range (FSR), 49,
  59–60, 63, 75, 83
front end of the line (FEOL), 225
FSR, *see* free-spectrum range
FWM, *see* four-wave mixing

gain, 199, 211, 262, 269, 280,
  282–83, 292, 296, 299, 302–3,
  305–6, 309, 320, 322

frequency-dependent, 278–79
internal multiplication, 199
material, 319
modal, 283
positive, 302, 306
steady-state, 285
graded index (GRIN), 97–98,
  106–8
grating couplers, 6, 96–98, 108–15,
  117
  high-efficiency, 113
  horizontal, 117
  long, 111–12
  out-of-plane, 112
  tapered, 271
  vertical, 114–15, 117
gratings, 49, 108–16, 156–57
  aperiodic, 118
  diffraction, 109, 112–13
  first-order, 308
  second-order, 111
GRIN, *see* graded index
guided mode, 26, 29–31, 36–37,
  63, 68, 111, 133, 139

HBTs, *see* heterobipolar transistors
Heisenberg uncertainty principle,
  172
Helmholtz equations, 28, 31
Helmholtz scalar equations, 26
heterobipolar transistors (HBTs),
  181

ICP, *see* inductively coupled plasma
indirect bandgaps, 10, 167, 172,
  186, 217, 226, 230, 236, 252,
  319
  low-energy, 170
indirect semiconductors, 11, 17,
  167, 169–71, 173–74, 185,
  194, 271, 273, 284–85,
  287–89, 291, 294, 297–98, 303
indirect transitions, 10, 16, 173,
  190, 202, 226, 252, 259,
  264–65, 318

lowest, 189, 203, 252
weak, 18, 260
inductively coupled plasma (ICP),
74, 76–77, 100, 106–7, 116,
131–33, 140
insertion loss, 70, 88, 96, 98–100,
102–3, 116, 228–29
additive, 116
theoretical, 105
intensity modulation, 226–28,
230–31, 234, 236, 241, 247
interference, 5, 82, 158–59, 226,
236, 242, 321
constructive, 88, 110, 113, 152,
245
destructive, 82, 245
internal quantum efficiency (IQE),
195, 206, 299–301
IQE, *see* internal quantum
efficiency

junction device, 256
junction diode, 299
junction photodiode, 191
junctions, 43, 169, 175, 193–96,
217, 236, 238, 256, 258, 260,
271, 299
abrupt, 199, 212–13, 215
biased PN, 240
frontside, 184
metal-semiconductor, 195
reverse-biased, 169

Keldysh effect, 208–9, 226, 230–31
Kerr coefficient, 85–86
Kerr effect, 85, 229–30, 234
Kerr self-phase modulation scale,
161
Kramers–Kronig dispersion
relation, 15, 230, 233

lasers, 135–36, 243, 245, 269–78,
280, 282–84, 286, 288, 290,
292, 300–302, 306, 308–12,
318, 321

diode, 271, 278
direct bandgap, 280
external, 5
external clock, 5
frequency-tunable, 79
group III/V, 74, 308–9
group IV heterostructure, 309
high-efficiency, 270
high-performance, 5
high-power, 270
hybrid, 310
infrared, 303
monolithically integrated,
269–70
monolithic heterostructure, 310
on-chip, 309
optically stimulated, 266, 273,
275, 277, 310
pumping, 275
real-world, 292
top-lying thinned, 308
tunable, 141
well-developed III/IV, 308
lasing, 6, 136, 276, 303, 306, 311,
319–20
lattice mismatch, 6, 18, 175, 190,
203, 217–18, 254, 261–62,
266, 300, 319
lattices, 9–10, 115, 126, 137, 169,
190, 251
air-hole triangular, 138
reciprocal, 127
stable, 9
LEDs, *see* light-emitting diodes
light coupling, 96, 98–99, 105,
108–11, 113, 115, 117, 141,
226, 270, 308
direct, 100
high-efficiency, 105, 117
horizontal, 115
vertical, 98, 114–15
light-emitting diodes (LEDs), 176,
255, 273
light waves, 27, 30–31, 35–36, 39,
146, 227, 308

Index

lithography, 40, 73–74, 109,
131–32, 219, 320
  deep UV, 131
  e-beam, 131, 133
  gray tone, 101
  i-line, 74
  nanoimprint, 89
  optical, 131
  polishing grayscale, 106
  ultraviolet grayscale, 100
logic gates, 50, 87, 89
  all-optical, 87
  optical, 87
Lorentz function, 142
loss, 21–22, 27, 38–41, 43–44, 72,
75, 89, 96, 101, 129–30, 133,
149, 303, 306, 309
  continuous, 302
  extinction, 292, 306
  internal, 68, 72
  leakage, 116
  modal, 292
  radiation, 40, 43, 69, 73
  reflection, 100
  refraction, 100
low-pressure chemical vapor
deposition (LPCVD), 78
LPCVD, *see* low-pressure chemical
vapor deposition

Mach–Zehnder interferometer
(MZI), 82, 89, 159, 226, 228,
236, 241–43
matter–light interaction, 179
  weak, 226
Maxwell's equations, 14, 22–24,
46, 105, 125–27, 153
MBE, *see* molecular beam epitaxy
metal-insulator-semiconductor
(MIS), 195
metal-oxide-silicon (MOS), 182,
195–98, 236, 239–40
metal-semiconductor-metal
(MSM), 196

microcavities, 130, 137–40, 142
  active optical photonic crystal,
137
  coupled, 139–40
  silicon-based PC, 137
  single-defect, 136
  symmetric, 138
microelectronics, 1, 3, 6, 21, 73–74,
89, 95, 105, 109, 114, 117,
202, 216, 320
microring resonators (MRRs),
49–60, 62, 64, 66, 68–90, 92,
146–47, 150, 154, 156, 236,
239, 243–46
  cascaded, 63, 161
  four-port, 54
  integrated, 49
  multiple, 49
  optical, 244
  series-coupled, 55–56, 88
  series-coupled double, 84
  single, 56, 149
  stable, 89
microrings, 51, 54, 56, 58, 61, 63,
68–69, 74, 80, 82–83, 88–89,
154, 246–47
  coupled, 82
  multiple, 52, 55, 83
  parallel-coupled, 52, 56–57
  series-coupled, 52
  structured, 152
mid-infrared (MIR), 180, 183, 200,
216–18
MIR, *see* mid-infrared
MIS, *see* metal-insulator-
semiconductor
MIS photodetectors, 197, 201, 211
MIS photodiodes, 197–98
MMI, *see* multimode
interferometer
mode mismatch, 96, 98, 102–3,
105
  horizontal, 116
  optical, 99

modulators, 50, 89, 138, 225–30, 232, 234, 236, 238, 240, 242, 244–47, 250, 270–71, 321
  electroabsorption, 226, 230, 265
  higher-frequency, 246
  high-speed, 230
  internal, 5
  laser light, 212
  optical, 85, 147, 232
  silicon-based, 243
molecular beam epitaxy (MBE), 215, 262
monolithic integration, 3, 100, 209, 269–70, 272, 308, 310, 317, 320
  complete, 317
  low-cost, 95
MOS, *see* metal-oxide-silicon
MQWs, *see* multiple quantum wells
MRRs, *see* microring resonators
MSM, *see* metal-semiconductor-metal
multimode interferometer (MMI), 70
multiple quantum wells (MQWs), 251, 255–56, 283
MZI, *see* Mach–Zehnder interferometer

nanophotonic waveguide, 96, 98–99, 106, 108–9, 115
n-doping, 193, 293, 297–98, 302, 305
near infrared (NIR), 181, 185, 190, 200–203
net gain, 302, 306, 308–9, 320
  positive, 276, 299, 301, 308
NIR, *see* near infrared
nonequilibrium, 168, 218, 237, 252, 262, 275, 278–80, 292
nonlinearity, 80, 85, 87, 161
nonradiative recombination, 168, 283, 299–302, 318–19

OEICs, *see* optoelectronics integrated circuits
optical absorption, 17, 72, 86, 230, 240, 259
optical buffers, 64, 81, 88–89, 102, 145
optical coupling, 96
optical devices, 50, 74, 135, 318
  compact, 49
  nonlinear, 85, 145
optical gain, 271, 275, 277, 279, 285, 306, 309, 318
optical loss, 50, 70, 306
  additive, 115
  high, 115
optical properties, 6, 9–10, 12, 14, 16, 18, 50, 79, 95, 172, 218, 254, 271, 304
optical spectrum analyzer (OSA), 79
optical waveguides, 1, 22, 24, 44, 74, 88, 100, 133, 148
optoelectronics, 1, 95, 181, 202, 214
optoelectronics integrated circuits (OEICs), 95–96, 101, 103, 108–9, 114
OSA, *see* optical spectrum analyzer
out-of-band rejection, 83–84

passive pixel sensor (PPS), 182
Payne–Lacey theory, 40
PBGs, *see* photonic bandgaps
PCs, *see* photonic crystals
  mid-infrared, 131
  silicon-based, 124, 138, 143
PCWs, *see* photonic crystal waveguides
  low-loss silicon, 159
  slot, 161
  slow-light propagation in, 156–57
p-doping, 193, 298

PECVD, *see* plasma-enhanced
 chemical vapor deposition
perfectly matched layer (PML), 128
phase modulation, 226–28, 230,
 236, 241, 247
phase shift, 51–52, 58, 147, 150,
 152, 243
 minor, 60
 single-pass, 149
phonons, 15–17, 86, 167–68, 170,
 172–73, 195, 230, 300
 folded, 272
photocurrents, 180, 194, 198–99,
 207–8, 217, 258, 260
photocurrent spectroscopy, 17
photodetectors, 180, 190, 202–3,
 211, 215, 217–18, 266, 271,
 273, 317
 first high-performance Ge-on-Si,
 202
 germanium, 5
 infrared, 211
 lateral, 219
 lateral incidence waveguide, 218
 resonance-cavity-enhanced, 216
 silicon-based, 201
 slow, 184
 small-area, 218
 vertical, 219
 vertical incidence GeSn, 218
photodiodes, 180, 182, 192, 198,
 206–9, 216–18, 270
 unitravelling carrier, 217
photoluminescence (PL), 168,
 172–74, 187–88, 259–60,
 271–72
photonic bandgaps (PBGs),
 123–24, 126–27, 129–30, 137,
 142, 157
photonic crystals (PCs), 6, 123–44,
 146, 156–59, 161, 322
photonic crystal waveguides
 (PCWs), 146, 156–61, 235–36
 slow-light, 160

photonic devices, 39, 41, 74, 117,
 175, 219, 244, 320
photonic integrated circuits (PICs),
 95–96, 109, 143, 320
photonics, 2–4, 6, 95, 179, 216,
 262, 269, 308, 310, 320
 group IV, 4
 high-performance, 143
 integrated, 49, 80, 86, 90
 ultracompact, 159
photonic systems, 21, 46, 169, 247,
 308–9
photons, 15, 123, 167–68, 170,
 185, 190, 230, 269, 273–75,
 279, 282
photoresists, 105, 131, 180
PICs, *see* photonic integrated
 circuits
PL, *see* photoluminescence
planar wave expansion (PWE),
 126–27, 139
planar waveguides, 21, 100, 105,
 182
 sandwiched, 130
Planck's constant, 201, 304
plasma dispersion effect, 226–27,
 232, 236, 242–43, 247
plasma-enhanced chemical vapor
 deposition (PECVD), 78–79,
 116
PML, *see* perfectly matched layer
Poisson number, 18
Poisson ratio, 264
polarization, 13, 22, 25, 38, 44–46,
 69, 79, 82, 89–90, 96–98,
 103–4, 106, 108, 116–17,
 141–42
PPS, *see* passive pixel sensor
prism couplers, 96–97, 105–7, 117
propagation constant, 22, 26–31,
 36, 72, 146, 228
propagation loss, 39, 41–43, 78,
 229, 239, 242
 optical, 78
 single-mode, 38
PWE, *see* planar wave expansion

QCSE, *see* quantum-confined Stark effect
  strong, 236
QDIPs, *see* quantum dot infrared photodetectors
QDs, *see* quantum dots
quantum-confined Stark effect (QCSE), 226–27, 235–36, 247
quantum dot infrared photodetectors (QDIPs), 211
quantum dots (QDs), 146, 169, 172, 200, 211, 218, 310
quantum efficiency, 168, 180–81, 195, 206, 209, 299, 301–2
quantum wells (QWs), 200, 211, 218, 226, 231, 235–36, 255, 263, 281–84
QWs, *see* quantum wells

radiation, 26–27, 39, 96, 117, 137, 140, 149, 167–68, 273, 275
  far-field, 137
  monochromatic, 109
radio frequency (RF), 77, 132, 181
Raleigh scattering theory, 73
Raman effect, 86
Raman lasers, 81, 86–87
Raman spectroscopy, 272
  polarized, 319
RC, *see* resistance-capacitance
recombination, 168–70, 194, 257–58, 299–302, 318, 322
  defect-dominated, 205
  direct, 300
  tunneling-assisted, 257
refractive index, 14, 24, 44–46, 85, 96, 98, 107, 115, 146–47, 156–57, 225–26, 228–39, 241–45, 247, 275–76
resistance-capacitance (RC), 2, 214, 271
resonance, 136, 138, 140, 151, 155, 216, 243–45, 276

resonators, 6, 43, 49, 61, 66, 74, 80, 85, 88, 148, 152–53, 156, 228, 236, 243–47
  56-microring, 154
  adjacent, 148
  all-pass, 82
  chain of, 148
  disc, 226
  microdisc, 243, 246
  microring/microdisc, 228
  neighboring, 153
  racetrack, 52
  ring, 226, 246
  spaced optical, 148
  standing-wave, 51
  tensile-strained microdisc GeSn, 311
  traveling-wave, 49, 51
reverse bias, 191–92, 207–8, 215, 238, 247
RF, *see* radio frequency
rib waveguides, 22, 30–38, 43–46, 68–69, 71–72, 117, 237
  etched, 38
  large, 35–36, 38
  multimode, 35
  shallow-etched, 43, 69
  single-mode, 100
  small-cross-section, 38
  submicron, 96
ridge waveguides, 2–3, 43, 71, 137, 141, 276
  tapered, 160
rings, 89, 155
  cascaded double, 63
  cascaded triple, 63
  coupled, 63
  coupled double, 63

scanning electron microscopy (SEM), 141, 160
scattering, 22, 27, 39–40, 77, 149, 276
  electron-photon, 304

intervalley, 284
intervalley electron, 284
optical, 73
stimulated Raman, 85
scattering losses, 40, 72–73, 101
random, 103
sidewall, 41
SCH, *see* separate-confinement
heterostructure
Schottky approximation, 194
Schottky barrier height, 196
Schottky contacts, 195–96, 256
Schottky photodiodes, 196
SCISSORs, *see* side-coupled
integrated sequence of spaced
optical resonators
56-microring, 155
balanced, 155
SEM, *see* scanning electron
microscopy
semiconductor heterostructures,
2, 4, 280
semiconductor lasers, 136, 273,
292
indirect, 287
stimulated, 299
semiconductors, 1–2, 4, 9–10, 13,
15, 17, 168–69, 171, 184–85,
190–93, 195–99, 211, 271,
273, 275–79
bare, 191
diamond-type, 186
doped, 175
group-IV, 229
high-doped, 195, 213
indirect-gap, 272
low-doped, 194
noncentrosymmetric, 235
n-type, 300
small-bandgap, 180, 215–16
separate-confinement
heterostructure (SCH), 282–83
Shockley–Read–Hall (SRH),
300–302

side-coupled integrated sequence
of spaced optical resonators
(SCISSORs), 148–49, 151–52,
154
sidewalls, 40, 73, 76, 102–3
silicon-based photonics, 4–6,
218–19, 265–66, 269–70, 308,
310, 317, 321
silicon-based waveguide, 146
silicon-on-insulator (SOI), 2–3, 21,
28, 49–50, 77, 81, 95, 126, 131,
146, 156, 174, 307
photonic, 272
silicon photonics, 1, 5, 89, 161,
272, 317
silicon waveguides, 3, 21, 38–39,
44–45, 103, 107, 116, 225,
235, 243–44
wide, 308
single-mode conditions, 22, 30–36,
38, 46, 68, 102
single-mode fibers, 35, 79, 96, 100,
104, 112–14
single-mode operation, 30–31, 35
single-mode waveguides, 3, 29, 35,
96, 102
single quantum well (SQW),
282–84
slab waveguides, 22–32, 46, 71–72
asymmetric, 73
multilayer, 30
silica three-layer dielectric, 30
slow light, 6, 66–67, 145–54,
156–62, 164
SOC, *see* system on chip
SOED, *see* spin-orbit energy
difference
SOI, *see* silicon-on-insulator
Soref's formula, 36–37
spin-orbit, 189, 253–54, 265, 289
spin-orbit energy difference
(SOED), 289
spot-size converters, 98–101, 103,
108–9

inversed tapered, 108
inverted tapered, 117
slot-waveguide, 104–5
vertical tapered, 99
SQW, *see* single quantum well
SRB, *see* strain-relaxed buffer
SRH, *see* Shockley–Read–Hall
SRS, *see* stimulated Raman
scattering
Stark effect, strong, 235
stimulated Raman scattering
(SRS), 85–86
stop-band, 62, 81, 134, 142, 157
strain, 19, 175, 189–90, 203, 216,
218, 235, 254–55, 263–66,
290, 292–93, 295–99, 301–2,
306–11, 318–19
elastic, 6
equal, 264
high, 311
in-plane, 265
mechanical, 319
residual, 18, 266
slight, 309
tensile elastic, 322
thermal mismatch, 251
uniaxial, 18–19, 264, 266, 298,
318
strain-relaxed buffer (SRB), 203
Stranski–Krastanov mode, 211
stress, 45–46, 69, 265
biaxial in-plane, 266
tensile, 175
strip waveguides, 30–35, 41,
43–45, 71
single-mode, 41–43
superlattices, 85, 169, 171–72,
271–72
system on chip (SOC), 181

tapered spot-size converters,
96–101, 106, 117
tapered structure waveguides, 98

TCO, *see* transparent conducting
oxide
TDs, *see* threading dislocations
TE, *see* transverse electric
tensile strain, 18–19, 175, 190,
216–18, 251, 266, 290–91,
293, 295, 297–98, 308, 318–19
high, 218, 290, 298, 309
large mechanical, 319
slight, 305, 309
strong, 297
thermo-optic effect, 80, 84, 154,
225, 227, 231–32, 240, 247
THG, *see* third-harmonic
generation
third-harmonic generation (THG),
160
threading dislocations (TDs),
151, 213, 217, 257, 319
undesired, 319
TM, *see* transverse magnetic
TPA, *see* two-photon absorption
transfer matrixes, 50, 52–53,
55–57, 149–51, 153
transitions, 10–11, 16, 167–68,
185, 187–89, 200–201,
211–12, 230, 251–52, 273–74,
276, 278–79, 287–88, 304–5,
318
transparent conducting oxide
(TCO), 197
transverse electric (TE), 25–30,
32–33, 38, 40, 42–45, 69,
82, 102–3, 105, 126, 128–9,
140–42, 240
transverse magnetic (TM), 25–26,
28–33, 44–45, 69, 82, 102–3,
105, 127, 129–30, 142, 240
two-photon absorption (TPA), 85,
87

ultraviolet (UV), 168, 185, 190,
200–201
UV, *see* ultraviolet

vacancies, 262, 300, 318
vacuum, 14–15, 30, 51, 71, 77, 146
valence bands, 10, 15, 185,
253–54, 263, 265, 275,
278–80, 285, 287, 289–92,
295, 298–99, 302, 306
direct, 289
light hole, 19
occupied, 13
strain-split, 292
valence sub-bands, 265, 290, 320
valleys, 187, 264–65, 307
Vernier effect, 63
Vernier effect of cascaded
microrings, 64
vertical incidence, 181–82, 209,
216
very-large-scale-integrated (VLSI),
245
VIS, *see* visible
visible (VIS), 3, 15, 169, 172, 185,
190, 200–203, 271, 286, 301,
303, 306
VLSI, *see* very-large-scale-
integrated

wafer, 2, 74, 100, 102, 108, 129,
131–32, 140–41, 239, 269–70,
272
waveguides, 1–6, 21–22, 26–33,
36–41, 43–46, 68–74, 95–99,
102–3, 105–9, 111, 113–17,
138–42, 210–11, 225–26, 238
coupled, 70
disc, 246
drop, 149–50
electrical, 183
fiber, 116
input, 80, 149–50, 245
input/output, 241
microring/straight, 69
multimode, 37
nanowire, 46
parallel, 138

photonic, 35
photonic-wire, 155
racetrack, 245
ring, 245–46
shared, 272
side-coupled, 149
slot, 86, 104–5
small-dimension, 108, 116
straight, 43, 52, 69–70, 245
submicron, 40, 68–69, 76, 97–98
thick, 113
transparent, 225
wide, 112
waveguide structures, 3–4, 31, 68,
95, 100, 106, 115, 156, 175,
235, 272
slot, 99
wavelength, 3–6, 33, 71, 74–75,
84–86, 110–11, 150–51,
182–85, 217–18, 230–33,
243–44, 246–47, 251–52, 306,
321–22
wavelength division multiplexing
(WDM), 49, 61, 82, 247, 308
wavelengths
absorption, 321
cut-off, 266
direct-bandgap-cutoff, 216
fixed, 243
higher, 251
larger, 183
long, 230, 235
lower, 175
millimeter, 5
operating, 33, 235, 247
optical, 51, 73
resonance, 65
resonant, 60, 67, 87, 149
short, 86
special, 82
vacuum, 14, 110
wave numbers, 15, 30, 146, 159,
167–68, 170, 172, 288

waves, 25–27, 42, 146, 158, 209–10
  continuous, 302
  diffracted, 111
  direct upward, 112
  downward-propagating, 112–13
  electromagnetic, 14, 22–24, 26, 145
  guided, 106, 111
  incident, 111
  incoming, 111
  optical, 282–83
  oscillatory, 26
  planar, 126
  reflected, 113
  single-frequency, 23
  standing, 158
  transmitted, 112
  upward-radiated, 113
wave vectors, 10, 13–15, 110–11, 138, 185–86, 278, 297
WDM, *see* wavelength division multiplexing

XNOR, *see* exclusive NOR

zero bias, 193, 207–8, 215, 247–48, 321
zincblende, 9–11
zone edges, 171–72, 186
zone folding, 171–72